# The Diffusion of Political Innovation: *From Eastern Europe to the Soviet Union*

ZVI Y. GITELMAN
*University of Michigan*

SAGE PUBLICATIONS / Beverly Hills / London

Copyright © 1972 by Sage Publications, Inc.

Printed in the United States of America

All rights reserved. No part of this book may be reproduced
or utilized in any form or by any means, electronic or mechanical,
including photocopying, recording, or by any
information storage and retrieval system, without permission in writing
from the publisher.

For information address:

SAGE PUBLICATIONS, INC.
275 South Beverly Drive
Beverly Hills, California 90212

SAGE PUBLICATIONS LTD
St George's House / 44 Hatton Garden
London E C 1

International Standard Book Number 0-8039-0145-3

Library of Congress Catalog Card No. 72-77771

FIRST PRINTING

# CONTENTS

An Inventory of General Propositions on the Diffusion of
    Innovation . . . . . . . . . . . . . . . . . . . .   11

Agents, Clients, and the Tactics of Innovation Proposal . . . . .   19

Some Propositions on the Diffusion of Political Innovation from
    Eastern Europe to the Soviet Union . . . . . . . . . .   22

Rules of the Game . . . . . . . . . . . . . . . . . . .   29

The Diffusion of Political Innovation: Three Case Studies . . . .   33
        Yugoslavia . . . . . . . . . . . . . . . . . . .   34
        Czechoslovakia . . . . . . . . . . . . . . . . .   37
        Hungary . . . . . . . . . . . . . . . . . . . .   41

Hungarian Innovations and Soviet Domestic Politics . . . . . .   49

Innovation and Political Change in Communist East Europe . . .   53

    NOTES . . . . . . . . . . . . . . . . . . . . . .   54

# The Diffusion of Political Innovation: *From Eastern Europe to the Soviet Union*

ZVI Y. GITELMAN
*University of Michigan*

**It is widely acknowledged** that since the death of Josef Stalin the nature of the relationship between the USSR and the various states of Eastern Europe has undergone a profound and dramatic alteration. While that relationship is still in the process of definition by the protagonists themselves, it may be asserted safely that new forms of dependence and interdependence among the various elements in the Soviet-East European entity have been evolved. It is generally acknowledged that in the Stalinist era the diffusion of political innovation and the transmission of political messages proceeded overwhelmingly from the USSR to Eastern Europe.

Today the picture is a mixed one. On the one hand, the conventional wisdom seems to be that "It is realistic to assume . . . that no great change is possible in Eastern Europe without corresponding change in Russia itself."[1] On the other hand, East European political systems are often seen as change agents influencing the Soviet system, and generally in a "liberalizing" direction. "Eastern Europe itself has long been recognized as a kind of ideological antechamber to the USSR. What is permitted in Eastern Europe becomes politically available, so to speak, inside the Soviet Union."[2] Since East Europeans seem to have been more adventurous, imaginative, innovative, and heterodox than their Soviet counterparts in

---

AUTHOR'S NOTE: *An earlier draft of this essay was presented to the Conference on the Influence of Eastern Europe and Western Areas of the USSR on Soviet Society, sponsored by the Center for Russian and East European Studies, the University of Michigan, May 1970. The research assistance of Judy Donald and Patricia Kolb, as well as the comments of Professors Melvin Croan, Steven M. Goldstein, and William Zimmerman, are gratefully acknowledged.*

such fields as economics, sociology, psychology, politics, and literature, the East European "antechamber" has introduced "liberal" ideas, and in some cases, practices, into the Soviet system.

It is not the purpose of this essay to test competing hypotheses about the direction of innovation flow; its aim is merely to provide an anatomy of the innovation process and its effects. To the hypotheses about innovation flow, we must add the possibility that the interdependent relationship of the Soviet Union and Eastern Europe may also result in a less visible, but no less real, conservatizing effect on the USSR. The USSR is burdened with the necessity of providing political leadership and policy cues for the East European states, and it must reckon with the fact that just as what is politically permitted in Eastern Europe becomes available in the USSR, so that which is politically legitimate in the USSR will be politically available to Eastern Europe. For this reason, since the late 1950s, at least, major doctrinal pronouncements in the Soviet Union have been made "with at least one ear cocked toward possible bloc reactions and domestic echoes of bloc problems."[3] Thus, it may be that Eastern Europe acts as a depressant on liberal innovation in the Soviet Union.

The potential Soviet innovator may reason that while a well-socialized and politically reliable Soviet citizenry can absorb significant policy and institutional innovations with no destabilizing consequences to the system, the less thoroughly socialized and less reliable populations of Eastern Europe might react in unpredictable and destabilizing ways to the introduction of such measures. The Soviet population, for whom Stalinism had been a more intense and prolonged experienced than for any other East European nation, was able to adjust to de-Stalinization in a more satisfactory way, from the point of view of the leadership, than the Polish and Hungarian populations, or even the Rumanians and Czechoslovaks. Similarly, it may well have been the case that the Soviet Union was prepared to make economic and political overtures to West Germany in the 1960s, but it was constrained by fear that the expansion of relations between West Germany and some of the other socialist states would weaken the socialist alliance and would have detrimental domestic political effects in Poland. Czechoslovakia, and perhaps some of the other socialist states as well.

Our judgment as to the general nature of the East European influence on the Soviet political system rests ultimately on our conception of the nature of the relationship of the USSR and Eastern Europe and on empirical investigation of influence flows between the entities. It would be very useful for heuristic purposes if we were able to characterize the USSR/Eastern Europe entity as analogous to a known entity or type of relationship. The latter could then serve as a model for the USSR/Eastern Europe, and we could study the process of innovation diffusion in the model in order to generate testable hypotheses about the same process in the entity we are analyzing. However, the USSR/Eastern Europe in the post-Stalin era seems to be a regional alliance of a very special—perhaps unique—type, and it is difficult to find a heuristically useful model for it. Therefore, we shall content ourselves with describing and analyzing some of the defining characteristics of this entity, as well as indicating the changes that have transformed the Soviet empire into the present entity.

It may be useful to regard USSR/Eastern Europe in the Stalinist era as an empire. According to Ghita Ionescu, an empire is characterized by three basic elements:

> (1) a political center animated by a historical mission of expansion; (2) religious or ideological coercion used to weld it into a single coordinated or expanding unit; (3) a sense of final purpose justifying it and inspiring its officers, soldiers and officials to transcend their individual role in the particular phase of development in which they find themselves. In the case of Stalin's Russia the three elements can be clearly seen.[4]

The imperial nature of the Soviet-East European relationship was manifested in complete political subordination of Eastern Europe to the Soviet Union, the methods of deviance control employed by Stalin, the economic relationships between the Soviet Union and the East European countries, the position and role of Soviet "advisers" and diplomats, as well as police functionaries, in Eastern Europe, the patrimonial leadership system, the narrow range of permissible political behavior, and, in general, the conformity to and imitation of the Soviet model that was forced upon all satrapies. To be sure,

> even in 1949-1953, years which are considered to have been the "golden age of unity," the influence of the objective specific conditions was apparent in the policy of the individual Communist and workers' parties, at least to some degree, although this fact was not part of the theory on the relations among the socialist states at that state of development.[5]

But despite variations from the norm, such as Poland's abstention from collectivization of agriculture, it can be fairly stated that the Soviet Union and Eastern Europe were involved in an essentially imperial relationship.

It was the Soviet leadership after Stalin which consciously set about to alter this relationship, and it was supported in this effort by some of

Stalin's former satraps who were eager to change their roles from that of patrimonial satraps at least to feudal vassals able to generate and command significant power resources of their own. The Soviet leadership, and particularly Khrushchev, changed both the theory and practice of Soviet-East European relations. Whereas, previously, coercion had been the usual way of repressing strains and controlling deviation on any major issue, Soviet leaders now entered bargaining relationships with East European leaders, though the coercive option remained available for use in extreme cases. "To bargain with subordinates implies some lessening of differences of status,"[6] and it is clear that this change in behavior was symptomatic of a larger change in the nature of the Soviet-East European relationship as a whole. R. V. Burks has written of changes in three types of linkages between the Soviet Union and the East European states:

(1) whereas under Stalin, leaders in Eastern Europe were "appointed and removed by the Kremlin much in the fashion of party cadres in the Soviet Union proper... Moscow no longer appoints and Moscow no longer removes." (Although, one might add, Moscow retains influence and perhaps even veto power over major appointments.)
(2) Policy formulation and implementation, once the exclusive privilege of the Soviet Union, is now a national responsibility, though it is in the interest of East European states to "follow the Soviet lead in all matters which do not affect them directly" since "Accretions to Soviet power and prestige in the international area generally reflect favorably on the kindred Communist polities of Eastern Europe."
(3) Ideology has become more important as a cohesive force holding the USSR and Eastern Europe together.[7]

Along with the diminution of coercion and the increased importance of ideology as cohesive forces, organizational devices and formal institutions, such as COMECON and the Warsaw Pact, replaced the highly informal and personal ties connecting Stalin with the East European satraps. Furthermore, whereas in the Stalinist system all units of the empire were isolated from each other to a great degree, so that each unit would form dependent and firm ties only with the Soviet center itself, in the present stage all sorts of genuine lateral ties have been formed among the East European states. They are now linked to each other, as well as to the Soviet Union, by a variety of formal instruments and informal devices. Moreover, whereas subnational units, such as professional groups, had been isolated and sealed off from their counterparts in other countries during the Stalin era, today these ties exist and seem to be growing in intensity and number. Thus, in some senses there is a higher degree of integration in USSR/Eastern Europe today than there was in 1949-1953. In fact, this integration creates a new type of political system, a "concentric" or "dual" one, whose characteristics and importance we shall explore later.

In somewhat more general terms, we can characterize the changes that have come about in the post-Stalin era as changes in the mix of prescriptive and restrictive messages emanating from the Soviet Union and transmitted to Eastern Europe, and as changes from consensual relations to some approximation of cooperative ones. The Soviet Union has ceased to issue, directly or by implication, detailed directives on domestic and foreign policy to its East European allies, but it continues to define the limits within which policy choices can be made, those limits having been broadened considerably since 1953. This shift from a prescriptive to a restrictive role is paralleled by a similar shift in the domestic activities of Soviet and East European governments, with the exception of Albania. Kenneth Jowitt has pointed out that the changing nature of relations among the socialist states of Europe can be understood as a change from consensual relations, as Irving Louis Horowitz defines them, to cooperative ones. Consensual relations involve

> shared perspectives, agreements on the rules of association and action, a common set of norms and values. [Cooperation] makes no demands on role uniformity but only upon procedural rules. Cooperation concerns the settlement of problems in terms which make possible the continuation of differences and even fundamental disagreements. ... Consensus is agreement on the content of behavior, while cooperation necessitates agreement only on the form of behavior. ... Cooperation concerns toleration of differences, while consensus demands abolition of these same differences.[8]

In brief, consensus relations demand conformity on means as well as ends, whereas in cooperative relations there is agreement on ends, and agreement to disagree on means. The fundamental difficulty in Soviet-East European relations is that while lip service is paid consistently to the shift from consensual to cooperative relations, very often counterdoctrines (such as the so-called "Brezhnev Doctrine") are enunciated which effectively vitiate the agreement to disagree on means, the rationale for these counterdoctrines often being that betrayal of goals, rather than pursuit of different means, has "objectively" taken place.[9] Certainly, the proclaimed change to the acceptance of "multiple paths to socialism" is frequently contradicted by inconsistencies and contradictions in behavior. The problem is that the Soviet Union has not committed itself decisively and consistently to cooperative relations with other socialist states. Since the nature of the relationship and the rules of the game are unsettled and sometimes unknown, the players must operate in an atmosphere of high uncertainty and risk. This makes life especially difficult for innovators and change agents, who cannot easily predict and anticipate Soviet reactions to their innovations.

If these are some of the ways the Soviet bloc has evolved, the question remains as to whether we can comprehend its nature sufficiently to discern consistent behavioral characteristics or to reason about it by analogy to a model. Most Soviet descriptions of USSR/Eastern Europe refer to it as a "socialist commonwealth" (sodruzhestvo), but since so much of Soviet writing on the subject does not separate the normative from the descriptive, treating what ought to be as what is, it is not very helpful for our purposes.[10] Western scholars emphasize the irregularity of intrabloc relations. "The old one-way relations have been replaced by an operational and loosely institutionalized sub-system harboring an atmosphere of mutual dependence and value-sharing."[11] Jan Triska argues that in the world Communist movement, including its East European component,

> There is no single state rule-making or policy-articulating organ, although there are contenders for the role. Rule implementation, conflict containment, and decision enforcement are other weak points in the system. The party-states (as well as the system as a whole) depend for their common endeavor on *agreement and persuasion,* which oscillates from various degrees of assertion to tentative consultative or advisory assistance, depending on the particular actors involved.[12]

In sum, the Soviet-East European alliance system is still very much in the process of self-definition, of working out in theory and in practice the fundamental regularities of its operation. For this reason, we cannot speak of a stable process of the diffusion of political innovation within the entity, but must content ourselves with observations on that process at work in several instances, and try to formulate some generalizations about the process in the awareness that the nature of the process will change as the nature of the entity itself changes. We will therefore examine some general propositions about innovation diffusion, point out some of the unique characteristics of innovation diffusion in Communist systems, and, finally, examine two cases which will illustrate the "rules of the game" of innovation diffusion.

# AN INVENTORY OF GENERAL PROPOSITIONS ON THE DIFFUSION OF INNOVATION

Some of the propositions about innovation diffusion have been selected from the literature on the subject in order to better understand the specifics of the process in USSR/Eastern Europe. This set of propositions does not purport to rank-order the most important findings about diffusion innovation, nor even to include them all. It is also not our purpose to test these propositions by the empirical materials. The propositions are designed merely as a checklist, useful for heightening our awareness of how the innovation diffusion process has been perceived and analyzed, and for illuminating the particular process we are studying.

There are many definitions of innovation which vary largely according to the context in which the innovation is studied. In his discussion of innovation in organizations, Lawrence Mohr defines innovation as "The successful introduction into an applied situation of means or ends that are new to that situation. Invention implies bringing something new into being; innovation implies bringing something new into use."[13] For our present purposes we will define innovation as the development and implementation of a program or policy which is institutionalized and which is generally acknowledged to have systemic significance.

In order to persist and develop, political systems must be attentive and responsive to their environments, including the extranational environments.

> Without openness to new information from their environment... self-steering organizations are apt to cease to steer themselves and to behave rather like mere projectiles entirely ruled and driven by their past.... Every self-governing system must... remake its own memories and inner structure as it acts. These inner changes may be small or large at any particular step, but their cumulative effect is apt to be considerable.[14]

There may be a variety of obstacles to this learning process. Karl Deutsch focuses on "will," implying the desire not to learn, and "power," implying the ability not to have to do so.

> Will and power may easily lead to... self-destructive learning, for they may imply the over-evaluation of the past against the present and future, the over-evaluation of experiences acquired in a limited environment against the vastness of the universe around us; and the over-evaluation of present expectations against all possibilities of surprise, discovery, and change.[15]

Both the will and the power of the Soviet Union may impede its reception of inputs, including political innovations, from the East European states, especially since for historical and ideological reasons the Soviet Union perceives itself as a teacher and guide, rather than as a student and disciple, of other socialist states.

It is crucial to our understanding of the Soviet reception of innovative proposals emanating from Eastern Europe to be aware of the way in which the USSR establishes its criteria for selection and judgment of those proposals.

> The individual in a complex organization . . . does not deal directly with all the sources of information potentially available to him, nor does he evaluate every conceivable policy option. In place of the debilitating confusion of reality he creates his own abstract, highly simplified world containing only a few major variables. In order to achieve this manageable simplicity he adopts a set of decision rules or standard criteria for judgment which remain fairly stable over time and which guide him in choosing among sources of information and advice. A decision maker decides both where to look for cues and information and how to choose among alternatives according to his decision rules; these rules also embody the current goals and aspirations of his organization, or the values which the organization is designed to advance and protect. Hence, if we wish to predict the decision maker's behavior, we should try to discover these rules of thumb . . . which shape his judgment. His choices could then be explained in terms of the alternatives he considers, his knowledge of each alternative, the sources of his knowledge, and the standard decision rules he applies in cases of this kind.[16]

If we may extrapolate from the individual in an organization to decision makers in a political system, we can see the relevance and importance of delineating the "rules of thumb" Soviet political decision makers use when considering innovative political alternatives. We should be able to predict with greater accuracy which innovative proposals stand a better chance than others in gaining Soviet attention, and which are more likely to be adopted, accepted, or tolerated. We shall explore this subject in the third section of the essay. For the moment, let us enumerate several propositions about the diffusion of innovation which are of a less general and abstract nature.

(1) "Several acculturation studies seem to show that the tangible aspects of any culture are more readily diffused than are purely ideological or behavioral aspects."[17] Tangible objects are easier to demonstrate and copy than abstract ideas. Secondly, "The advantage or disadvantage of one thing over another is more obvious than the advantage or disadvantage of

one institution over another because the potentialities of a thing are more closely related to its physical properties. . . . The real point here is that knives are judged by the same standards, whereas religious and marital customs are not."[18] It is probably easier for the Soviet Union to borrow purely technological innovations than it is to adopt new political ideas or institutions, even when they are proposed by kindred socialist states, though some technological innovations may have undesirable "spillover" effects into the political and social arenas.

(2) Innovations should demonstrate: (a) *relative advantage* over the idea or institution superseded; (b) *compatibility* with existing values and past experiences of the potential adopter; (c) *"divisibility* . . . the degree to which an innovation may be tried on a limited basis."[19] These three requirements are especially important in regard to the Soviet Union. Soviet decision makers, perhaps to a greater degree than their Western counterparts, are conditioned to assume that they live in the best of all possible worlds, and in order to persuade them to adopt a political innovation, its relative advantages must be clearly demonstrated. Secondly, since the Soviet system is an explicitly ideological one and is very conscious of its values and historical experiences, compatibility with Marxist-Leninist ideology, as defined by Soviet leaders, is a sine qua non for any proposed innovation. The principle of divisibility, which holds that "New ideas that can be tried on the installment plan will generally be adopted more rapidly than innovations that are not divisible,"[20] is of great importance in the context of USSR/Eastern Europe. The Soviet Union may be willing to experiment with economic reform, but it did so by gradually increasing the number of enterprises operating under the new system, rather than by introducing the reform across the board at a single time, as did the Czechoslovaks and Hungarians. The Soviet leadership would be more reluctant to introduce a sweeping Party reform, since that kind of change might have a greater "spillover" effect into many areas of Soviet life than would controlled economic reform. One could also conceive of the possibility that the USSR might want to introduce a new type of relationship with one or two of its East European allies, while maintaining existing relationships with others. This kind of change would not be easily divisible, and so it may be that the necessity to introduce the change, say, from consensual to cooperative relations, to *all* inter-socialist relationships effectively prevents the USSR from introducing such changes in *any* of them.

(3) The chances for adoption of an innovation are enhanced if that innovation has been successfully adopted in another similar setting. Jack Walker's study of the diffusion of innovation among American states led

him to conclude that "the likelihood of a state adopting a new program is higher if other states have already adopted the idea. The likelihood becomes higher still if the innovation has been adopted by a state viewed by key decision makers as a point of legitimate comparison."[21] The Soviet Union would be more likely to adopt an innovation that had been tested successfully in an East European state, particularly one which the USSR could view as being somewhat analogous to the USSR in important ways. Clearly, no East European state compares with the Soviet Union in size and world power, but it is likely that the USSR would be more interested in certain kinds of experiences in the more developed socialist countries than in the experiences of Albania, Rumania, or Bulgaria.

(4) Some commentators on innovation diffusion and communication postulate that "at least some parts of the receiving system must be in highly unstable equilibrium, so that the very small amount of energy carrying the signal will be sufficient to start off a much larger process of change."[22] In a study of innovation in organizations, James Q. Wilson argues that "many organizations will adopt no major innovation unless there is a 'crisis'—an extreme change in conditions for which there is no adequate, programmed response."[23] It is possible that internal instability may promote a search by Soviet political leaders for alternative modes of political or economic organization, and it is undoubtedly true that it is precisely in times of leadership instability and power struggle that innovative policies and programs are introduced. These become weapons in the leadership struggle. The system may be most open to the introduction of new programs and policies at times of leadership struggle. On the other hand, such a situation, while it promotes internally generated innovation, may retard the acceptance of externally generated proposals for change, since there may also be a reluctance to ask for the aid of supposedly inferior units in resolving disputes in the superior unit. This question should be explored further by examining the use made of East European programs and policies in Soviet leadership disputes, or more generally, in times of instability in the Soviet system as a whole.

(5) "A voluntary association with broad, diffuse goals (typically associated with relatively low salience) will adapt more readily to environmental changes than will organizations with narrow, precisely stated goals (typically associated with high salience)."[24] Relative to other political systems, the USSR has narrow, ideologically defined, precisely stated goals which have high salience both within and without the system. Since these goals are relatively specific and there is a high awareness of them, all innovations and changes must be justified in terms

of those goals. This imposes an additional burden upon the East European change agent, his Soviet linkage group or actor, and Soviet decision makers.

(6) "A generalization supported by many studies is that impersonal information sources are most important at the awareness stage, and personal sources are most important at the evaluation stage in the adoption process."[25] Soviet decision makers may become aware of East European innovations through publications and other impersonal sources, but when decision makers must evaluate the innovations, personal relations among leaders in Eastern Europe and the USSR and personal persuasion becomes important. One advantage that East German and Hungarian innovators may enjoy is that Ulbricht or Honecker and Kadar seem to have the confidence and respect of the Soviet politburo. This was not true of either the Novotný or Dubček leaderships in Czechoslovakia.

(7) Another "psychological" variable are the characteristics of the linkage groups which transmit innovation from one system to another. First, "A linkage group becomes much more susceptible to the inputs from abroad if its ties to the domestic system are weakened—if it is, for instance a segregated or discriminated minority, or if it is an economic class or social class which is disadvantaged or alienated."[26] This might apply to national minority groups, such as Ukrainians or Jews in the USSR, and could also include economic reformers, political nonconformists, dissident intellectuals, and other "non-establishment types." Of course, while these groups may be more receptive to assuming the role of linkage agent, they are less credible and trustworthy to the recipient of the innovative stimulus. If these groups could, in turn, persuade other, more powerful and prestigious groups to join with them in the attempt to convince the client to accept the innovation they may be able to overcome their low status disadvantage. In other words, in diffusing innovations from Eastern Europe to the Soviet Union it may often be necessary to reach two kinds of linkage groups: an initially receptive but low-status group which might, in turn, act as a linkage to a less receptive but more influential group.

(8) A third psychological factor to be considered is "that cognitive dissonance between a message and a past attitude is resolved by cutting down the message and retaining the attitude, if there is strong social support for the attitude."[27] Deutsch has found that this outweighs "by a factor of four or five to one [the] bandwagon effect." Support for traditional attitudes may not only be found; it may be mobilized as well. An innovative proposal emanating from Eastern Europe, or from within the Soviet system itself for that matter, will undoubtedly encounter strong resistance from "traditionalists" or conservatives, as well as from those who have a vested interest in maintaining the status quo.

(9) The psychological characteristics of the client are as important as those of the change agent and the linkage group. Certain individuals are more "open-minded" and receptive to change than others. Especially in Soviet-type systems, where individual leaders wield so much power, this variable can play an important, perhaps even decisive, role in determining the kind of reception given an innovative proposal. Clearly, Khrushchev's penchant for innovation was far greater than that of his successors who criticized him for being overly enamored of new, untested schemes.

(10) Information about the innovation proposed should be readily available to the client. "We can predict that an anticipated change will be resisted to the degree that the client system possesses little or incorrect knowledge about the change, has relatively little trust in the source of the change, and has comparatively low influence in controlling the nature and direction of the change."[28] The client should be made to feel sure about the consequences of the change. The change effort should be perceived as being self-motivated and voluntary. Finally, the change program "must include emotional and value elements as well as cognitive (informational) elements for successful implementation. It is doubtful that relying solely on rational persuasion (expert power) is sufficient."[29] The Soviet Union demands that proposed innovations, whether they are proposed for adoption in the USSR or merely for implementation in Eastern Europe, remain always under Party control, and that their consequences be fully explored. Naturally, it is desirable that the innovations have some expressive value and appeal. It also appears to be of great importance to the USSR to be able to present the innovation as being, in some form or other, self-motivated and certainly taken as a voluntary step, not imposed by any outside forces or by objective circumstances. All changes in Soviet policy and institutions are presented as "creative developments" of traditional values and forms, undertaken as voluntarist initiatives by a progressive system which is responsive to the need for change.

(11) Developing the idea of predictability of consequences further, it is important to note that there is an ineluctable element of spillover in the adoption of almost all innovations. Accepting one idea often means that supporting or complementing correlates must also be accepted.[30] Moreover,

> Sometimes the changes brought about simply "fade out" because there are no carefully worked out procedures to ensure coordination with other interacting parts of the system. In other cases, the changes have "backfired" and have had to be terminated because of their conflict with interface units.[31]

The Soviet Union cannot accept some types of economic reform since they do not seem to coordinate in an acceptable or desirable manner with "other interacting parts of the system"—the Communist Party *apparat*, for example. The Leninist formulation of "He who says A must say B" is especially true in a Soviet-type system where the various interacting parts are highly integrated with, and dependent on, each other.

(12) Several observers have pointed out the impediments to innovation diffusion and some of the strategies that a change agent can employ to overcome them. The anthropologist A. L. Kroeber points to three general checks on diffusion: lack of communication, resistance in recipient culture, and "displacement." The first is largely a technical problem, except in cases where technological and political impediments to communication are purposively contructed to protect the potential client from external ideas and influences. Stalin's attitude toward the Yugoslavs after 1948 and Soviet and East German measures taken to cut down the volume of communications emanating from Czechoslovakia in 1968 are familiar examples of this kind of behavior. Resistance, according to Kroeber, is usually due either to the presence of traits in the recipient system felt to be irreconcilable with the invading traits, or to "the presence of cultural habits functionally analogous to the new elements which results in a block. Coffee is unlikely to invade rapidly or successfully a nation addicted to tea drinking."[32] This suggests that there may be greater resistances to innovation diffusion among similar political systems—socialist ones, for example—because there are entrenched institutions which can easily interpret the innovative proposal as a threat to their own power, and because the innovation can more easily be regarded as superfluous—its equivalent is already present in the client system. If, on the other hand, the innovation is perceived as *not* functionally analogous to an existing institution, behavioral pattern, or value, it might be regarded as "irreconcilable" with or "hostile to" the existing defining characteristics of the system. One can envision at least four logical possibilities:

(a) if the innovation is functionally analogous to an element in the existing system, there is a high likelihood of resistance and rejection;
(b) if the innovation is not functionally analogous and is seen as conflicting with defining characteristics of the system, rejection is also likely;
(c) if innovation is neither functionally analogous nor seen as challenging to defining characteristics there is a greater chance for favorable consideration and adoption;

(d) if the innovation seems analogous to accompanying characteristics of the system there is a possibility of adoption, but a strong likelihood of resistance to adoption by the vested interests whose function is likely to be transformed or displaced by the innovative proposal. [33]

Karl Deutsch has explored a variety of ways in which a nation-state can reduce or stop the inflow of external inputs. It may count on the disappearance of the external source of inputs, reduce the linkage groups or institutions, cut off contact with the input source, make the domestic system more stable and, hence, more impervious to external inputs, or try to effect a change in the environment itself. The USSR has employed all these tactics against East European inputs at one time or another, not only to protect itself, but also to prevent innovative inputs from flowing from one East European country to another. (Stalin was undoubtedly correct in assuming that innovations would more easily pass from one East European state to another than from any of them to the USSR. While this essay deals with innovation diffusion from East Europe to the USSR, the intra-East European diffusion of political innovation is probably more frequent and perhaps more successful.) Stalin cut off contact with Yugoslavia, and reduced even potential linkage groups by the purges of so-called "Titoists," tried to shore up the domestic political system, and, in a way, counted on the disappearance or increasing irrelevance of the troublesome external threat. These tactics have also been used by Stalin's successors vis-à-vis Hungary in 1956 and Czechoslovakia in 1968. A less visible defense that the USSR has for resisting East European—or any other—inputs is the "objective" factor of size.

It is possible that the domestic system may get the same results by its sheer size. A very large country, very prosperous and with very strong holds upon its population, may be able to withstand even major impacts of foreign propaganda by tying its potential linkage groups so strongly to the domestic system that all foreign inputs become relatively insignificant. Similar effects can be obtained by multiplying and intensifying small group ties, even in a small country.... The ties of integration to the main system become so strong that any inputs from abroad to potential domestic groups remain quite ineffective. [34]

By interlocking the elites of all domestic groups with the Communist Party, and by assiduously cultivating a strong sense of "Soviet patriotism," the Soviet system has tied effectively its potential linkage groups, with a few exceptions, to the domestic system.

## AGENTS, CLIENTS, AND THE TACTICS OF INNOVATION PROPOSAL

These then are some of the conditions and tactics which militate against the diffusion of innovation from one political system to another. What are the counterstrategies that can be employed by the change agent? If East Europeans consciously or unconsciously play the role of change agents, how can they overcome Soviet resistance to innovation? We have already noted the necessity for a change proposal to be assimilable into the prevailing value system of the recipient, the necessity to have the recipient perceive a need for change, the importance of anticipating the consequences of change, and the wisdom of appealing to strategically placed groups or opinion leaders in trying to reach the final decision makers. Everett Rogers adds that

> Change agents should be more concerned with improving their clients' competence in evaluating new ideas and less with simply promoting innovations per se ... a long range program to change values may be a more appropriate strategy of attack for some change agents than just a "single innovation" approach to change.[35]

This is probably not a wise strategy to employ when dealing with a Communist system, one which has a profound commitment to a set of values which are not easily abandoned or revised. In fact, it might be best to convince the client that adoption of the innovation involves no change in values whatsoever. It is more advisable to calculate on the incremental erosion of present values and their transmutation over time than to attempt a frontal confrontation and critical assault on fundamental values. As Herbert Shepard points out, one way of innovating in a resisting organization is to conceal the innovation as much as possible.[36]

In a suggestive article on "non-conforming enclaves" in organizations, Ruth Leeds observes that such an "enclave" often presents its innovations as techniques designed to facilitate attainment of organizational goals, not as means of changing the goals.

> The commitment inspiring the non-conformists is frequently viewed as higher than that possessed by others in the organization ... likely to provoke conflict. The non-conforming enclave is further distinguished by an unorthodox atmosphere which permeates many aspects of its life.[37]

Yugoslavia in the early 1950s, Czechoslovakia in 1968, Albania since the early 1960s, and other nonconforming East European states have all presented themselves to the USSR and other socialist countries as affirming the fundamental goals of Marxism-Leninism and as being more seriously committed to the "transitive" goals of the Communist ideal than the "orthodox" states which have been diverted to the pursuit of "reflexive" goals and have departed from the true path.[38] Certainly, the defenders of orthodoxy have charged, and not without reason, that the general atmosphere prevailing in the deviating states is not in conformity with established practice and, what is essentially Soviet, tradition. The deviant or innovative enclaves respond, of course, that the defenders of orthodoxy have become so committed to their reflexive goals, that new proposals for transitive goal attainment can only be regarded with suspicion by those vested interests (Djilas' "new class") which respond to innovative proposals with "trained incapacity."[39]

Leeds mentions four basic ways in which the organization can deal with its nonconforming enclave:

(1) condemnation;
(2) avoidance;
(3) expulsion;
(4) protest absorption.

The first three tactics will work only if the enclave is weak. Condemnation will widen the rift by forcing a polarization of the issues (as has been the case in the USSR's dealings with Yugoslavia, Czechoslovakia, Poland, Rumania, as well as China). Avoidance might allow the enclave to grow rather than die out. Expulsion might be in effect a loss of resources (allies) which could have been channeled to serve organizational goals. Expulsion might also lead to the emergence of a rival structure—for example, an alternate "Communist International." Finally, if protest absorption is the strategy pursued, the nonconforming enclave may gain access to key positions and heavily influence or control the entire structure.[40]

In the protest absorption process the organization's leaders will confront the necessity to balance the demands of the nonconforming enclave against those of a "middle hierarchy" standing between the leaders and the enclave. The occupants of the middle-level roles are the ones most likely to exhibit trained incapacity and are most directly threatened by the innovative proposals of the nonconforming enclave. Soviet middle-level officials may be the most strenuous opponents of innovations which would upset Soviet institutional arrangements and programs. Analogously, the role of Gomulka and Ulbricht in pressing the Soviet leadership to halt the Czechoslovak experiment may be understood as that of something like middle-level officials trying to get top-level decision makers to reject decisively innovations with unacceptable implications for their own bailiwicks. If the middle-level resistance is overridden by the top leadership the enclave gains some autonomy to pursue its innovation.

> This is followed by several more rounds of obstruction by the middle hierarchy, unorthodox communication to the top by the non-conforming enclave, and a gradually increasing grant of resources, autonomy, and legitimacy to the enclave by the top hierarchy. With each round the enclave comes closer to approximating a new legitimate subunit. ... In exchange for autonomy and legitimacy from the top hierarchy, the enclave must agree to accept certain stabilizers ... mechanisms to insure the loyalty of the new unit to the organization and its conformity to organization regulations.[41]

These stabilizers include rules of conduct, a regular source of inputs, and agreement to limit the innovative activity to a particular sphere. As we shall see, all these have been operative in regard to the Hungarian economic reform. The Czechoslovaks' hints about the relevance of their innovations in all sorts of spheres and beyond their borders helped kill their chances for acceptance by the Soviet Union and its allies.

"In large measure, the significance of protest absorption for the organization as a whole depends upon the bearing which the enclave's cause has on the core policies and practices of the organization."[42] As we have stressed, the closer to the core of Communist doctrine and practice the innovation is, the more resistance it is likely to encounter by top-level decision makers, whether within an East European state or the USSR.

Finally,

> An organization which has had long experience with non-conformity, e.g., the Catholic Church, might institutionalize the rounds of protest absorption. ... If the adoption of protest absorption as a conscious organizational policy is carried out effectively, an organization will strengthen its ability to cope with non-conformity and to implement changes flowing upward from the bottom.[43]

The Stalinist experiences of rejecting and repressing nonconformity, rather than absorbing it, has slowed the learning process of the Soviet leadership. The Soviet leadership is just beginning to explore different ways of dealing with nonconformity, almost by definition involved in innovation, and it may be a long while before it develops a flexible response to innovation and nonconformity arising from Eastern Europe, or from any other quarter.

These then are some of the generalizations about innovation diffusion commonly found in the literature on the subject, and an attempt to suggest the utility and relevance of such generalizations for the study of the diffusion of political innovation from Eastern Europe to the Soviet Union.

## SOME PROPOSITIONS ON THE DIFFUSION OF POLITICAL INNOVATION FROM EASTERN EUROPE TO THE SOVIET UNION

It has been remarked that in the literature on innovation diffusion "The prevailing focus of attention is on the individual innovator ... not on the organizational setting in which innovation takes place."[44] In this section we propose to examine some of the factors specific to Communist systems ("the organizational setting"), especially the USSR, which cause them to deal with political innovation and its diffusion in a distinctive and probably unique way. After describing some of the salient features of the diffusion process from Eastern Europe to the USSR, we shall attempt to draw up some tentative "rules of the game" for the process, recognizing that the rules themselves are very much a matter of debate and are therefore not permanently fixed or universally agreed upon.

It has long been recognized that the Stalinist system, and particularly the "command economy" had a depressant effect on innovation, not only of a political nature but of a purely technological one as well.

> The immediate difficulty is that innovation, whether indigenous or borrowed, involves much labor and high risk, while the rewards that Stalinist central planning offers in return are inadequate to say the least. ... A totalitarian command economy is specially designed to prevent automatic, self-perpetuating sub-system processes, technological or otherwise. It is a basic principle that everything of basic importance must be done from above.[45]

Students of the Stalinist system are familiar with the dilemma of the industrial manager who cannot afford to innovate because the introduction of new processes or machinery would interrupt production schedules and the manager would fail to meet quarterly quotas, a failure with perhaps fatal consequences.

After the death of Stalin the Soviet Union was forced to take new departures in the fields of foreign policy and international relations doctrine, as well as in the economy. In both sectors there have been dramatic, highly visible new developments which have pressured Soviet leaders to fundamental reassessments and revisions: the advent of the nuclear era led to a change in Soviet global perspectives, and the increasingly inefficient and nonproductive command economy led to experimentation and partial dismantling. There has been no domestic political equivalent of the atom bomb, and, in contrast to some of the East European states, no dramatic political breakdown or crisis has occurred in the Soviet Union. For this reason, among others, there is no great and consistent pressure for political innovation in the USSR, though the Khrushchev era was distinguished (his successors might say marred) by several interesting political experiments and innovations. Czechoslovakia, Poland, Hungary, and the German Democratic Republic have been confronted with serious crises which have elicited responses attempting to establish regime authority on the basis of new political formulae.[46] The Soviet Union has faced difficult political problems since 1953, such as declining economic growth rates, nationality rebelliousness, intellectual dissent, failures in the international arena including the "space race," agricultural problems, and some generational problems, to single out a few of the most visible. The new *kolkhoz* charter, which essentially sanctioned the status quo, the failure to revise nationality policy, the uncertain role of the Communist Party, and the failure to proclaim a new constitution promised since 1963, point to a conservative choice to postpone systemic political changes, presumably in the belief that at the present stage system maintenance is more important than uncertain experiments designed to enhance systemic effectiveness. There is more "slack" in the Soviet system than in many of the East European ones because a longer and more effective socialization process, undoubted Soviet power and prestige resulting from impressive domestic and foreign achievements, and the isolation of the population from external influences have provided the USSR with a greater reservoir of legitimacy than have most other socialist countries in Europe. Finally, as was mentioned earlier, the dramatically different size of the Soviet Union enables it to deal with potential external inputs in a very different way from that in which smaller systems handle these inputs. It can turn its potential linkage groups inward much more easily than can, say, the Polish or Hungarian regimes.

The most obvious element distinguishing the innovation diffusion process among Communist systems from that obtaining within an organization is Marxist-Leninist ideology. One effect of the ideological nexus binding the socialist states together is to exacerbate differences of opinion among them and makes disputes harder to solve. "When opposing views or substantive issues are couched in doctrinal terms, each dispute in effect becomes two—one substantive and one ideological—and demands buttressed by claims of ideological purity do not lend themselves easily to compromise."[47] Brzezinski claims that ideological disputes are not easily resolved by compromise as financial or border disputes would be.

> Barring an outright split, the usual solutions tend to be more extreme: the agreement of one protagonist to become doctrinally silent, although not necessarily retracting his original views; or the doctrinal subordination of one to the other. A third solution, a mutual agreement to become silent, is not stable if one of the parties involved is the center, since a doctrinally oriented movement cannot remain doctrinally silent in the face of continually changing reality.[48]

Nish Jamgotch, who views ideology as the "irreducible unifying force" of the Communist bloc, argues that the very power of the Soviet Union as an "ideologically oriented great power ... must ultimately depend on its political transplants abroad."[49] While this seems to press the case for the importance of ideology too far, there is no doubt that ideology is an important component of Soviet-East European relations and that innovative proposals, even of a seemingly nonideological character, are very often evaluated in ideological terms, with the tacit or overt admission that the ideological effects of every innovation constitute one of its most important consequences.

A second, related, consideration is the claim that "In a monolithic alliance, if non-conforming behavior should be successful in one issue-area, there is a high probability that such behavior will 'spill over' into other issues areas."[50] It was the realization of this truth that prompted Stalin to take such elaborate prophylactic measures against Yugoslavia, and that explains much of the behavior of the Soviet Union, Poland, and the GDR in Czechoslovakia in 1968. But in at least one instance the alliance leaders opted for a different tactic of deviation control. Rather than "cracking down" directly on Rumania, the Soviet leadership has calculated correctly that it could afford to allow the "Rumanian deviation" because it would remain an isolated phenomenon, not successfully emulated in other socialist countries.[51] Here the tactic of partial isolation has been successfully used. The Rumanian Party leadership has demonstrated that it is capable of preventing the spillover of heresy from foreign policy to domestic affairs and this goes a long way to explaining Soviet tolerance of the Rumanian regime.

The doctrinal component of Communist relations also accentuates the usual relationship between deviance and innovation. Almost all innovators are ipso facto deviants since they depart from the prevailing norms. In

Communist systems departure from prevailing norms is assumed to be not merely deviant behavior proceeding from ignorance or lack of ability, but calculated nonconformity based on nefarious motivation and values. The worker who performed poorly in the Stalinist system was not merely a laggard but a "wrecker." The motives of the innovator are always probed deeply in order to discover the "real" reason for his advocacy of the innovation. This is reflected by the Communist rhetoric of "hiding behind leftist phrases," "concealing true motivations," being "unmasked," "posing as a friend of the labor movement," and the like. Especially in a system which has no place for political competition and where all legitimate alternatives are defined by the Party alone, nonsanctioned suggestions for innovation immediately raise questions about motivation and antisystemic attitudes. Even when the innovator denies deviation and strives very hard to prove his orthodoxy in deeds as well as in words, he is likely to be told that he is "objectively" a traitor, since by departing from official norms and even indirectly challenging authority he "objectively plays into the hands of the enemies of socialism." "Innovators are forced into a combative position" because "their novelties enter a social organization most of whose establishments are going concerns" and their innovations "enter as competitors or deprecators of one or another." [52] This general truth is magnified several times in the context of Communist systems and alliances.

Turning now to some structural considerations, we hypothesize that the type of leadership situation in the Soviet Union may directly affect the fate of political innovations, whether ideological or institutional, emanating from Eastern Europe. It has often been said of the present Soviet leadership that its general conservatism is partially explained by the fact that a coalition leadership cannot strike out in bold, new directions since it needs to gain consensus on major policy decisions, and consensus decisions tend to be agreements on the lowest common denominator. Radical policies and experimentation are potentially destabilizing and are therefore to be avoided. This means that innovations with system-destabilizing implications will not be entertained by a collective leadership. This is persuasive only if there is no one in the leadership who is looking precisely for a chance to destabilize the leadership itself. If a member of the coalition is seeking to enhance his own power at the expense of others, he may be very tempted to borrow and champion an innovation as a "new look" or better policy. Khrushchev's breakup of the post-Stalin leadership and his ascendancy to the role of single or foremost leader was achieved in exactly this way. Under the one-man-leadership system the leader may adopt innovations since he feels secure enough to experiment with them;

he may reject them if he feels that his position is so precarious that a disastrous experimental failure would seriously threaten or erode his power, and in some instances he may even be forced to adopt innovations in order to secure his position. Tito in 1950 and Khrushchev six years later used extensive institutional and ideological-policy innovations as weapons with which to fight off external and internal enemies and as tools to build more secure power bases for themselves.

Psychological as well as structural characteristics of leadership are important in determining the fate of an innovation proposal. Especially in systems where power is so highly centralized in the hands of a relative few, their psychological characteristics, insofar as they affect their political decisions and preferences, must be taken into account. To put it simplistically, we suggest that certain personality types are more likely to be receptive to innovation than others. There is differentiation among Communist leaders along these dimensions, just as there is among others. Some leaders tend to have more "open minds" than others, in the sense that while they too are committed to an ideology they hold to it within an open mind which retains the capacity for reconceptualizing the nature of problems and synthesizing the elements in a new way. Those with "closed minds" have only analytic capacities—they can break down problems and suggest specific solutions to component parts, but they cannot reintegrate the whole into a new synthesis. Without being so foolish as to attempt long-distance psychology, it is tempting to suggest that Antonin Novotný consistently exhibited the traits of a closed mind, whereas Alexander Dubček, a man of very similar background, training, and experience, displayed traits of the open mind, at least in 1967-1969.[53]

One element in the USSR's claim to primacy is the fact that it was the first, and for many years the only, socialist state in the world. Vernon Aspaturian comments that

> There is ... something absolute and irrevocable about chronological primacy. It can be neither reversed by historical repagination nor erased by ideological deracination. As the first party-state, the Soviet Union will continue to be the inspiration, direct or indirect, of all future Communist states, just as it survives as the common genealogical ancestor of all existing party-states.[54]

The chronological primary of the Soviet Union is a brick in the barrier against the penetration of East European innovations because the "Soviet model" serves as the standard by which all innovative proposals are judged. The undoubted long-term successes of the Soviet Union domestically and internationally buttress the conviction that the most efficacious political system in the social world is the Soviet one, and the proponents of that system look with skepticism at attempts to "improve" the model. Success itself may prevent further learning and development.[55] The "importance of being first" is sufficient compensation for any possible feelings of inferiority vis-à-vis the more "sophisticated" Poles, Czechoslovaks, or Hungarians. The conviction that the USSR is the socialist state closest to a Communist society implies that to borrow from less developed socialist states does not make historical good sense.

There are generally two ways in which East European ideological and institutional innovations make an impact on the Soviet political system: as the bearers of otherwise unacceptable non-socialist political ideas and mechanisms, and as inputs into debates among interest groups in the USSR. Since the loosening of the alliance and the dismantling of the empire, the role of Eastern Europe as a relevant factor in the domestic political process in the USSR has increased, and the increasing legitimacy and stature of the East European states, combined with the greater openness of the Soviet system, has made it possible for the East Europeans to serve as "transmission belts" for Western ideas. A prime example of this is the role of Polish sociology in the transformation of Soviet sociology into a discipline more closely approximating Western sociology.[56] Since empirical sociology was legitimate and officially sanctioned in one socialist country, it could be adopted by a second socialist country. What could not be borrowed directly from the West, could be adopted if it had a socialist imprimatur.

In his study of national-international linkages, James Rosenau delineates three basic types of linkages processes:

(1) "penetrative"—where members of one polity actually participate in the political process of another. "That is, they share with those in the pentrated polity the authority to allocate its values."
(2) "reactive"—"The actors who initiate the output do not participate in the allocative activities of those who experience the input, but the behavior of the latter is nevertheless a response to behavior undertaken by the former."
(3) "emulative"—this corresponds to "The so-called 'diffusion' or 'demonstration' effect whereby political activities in one country are perceived and emulated in another."[57]

In the Stalinist era the Soviet-East European linkages were both penetrative and emulative. At present, the penetrative linkage has been reduced, though Soviet troops still remain in some East European countries and, as is obviously the case in regard to Czechoslovakia, they

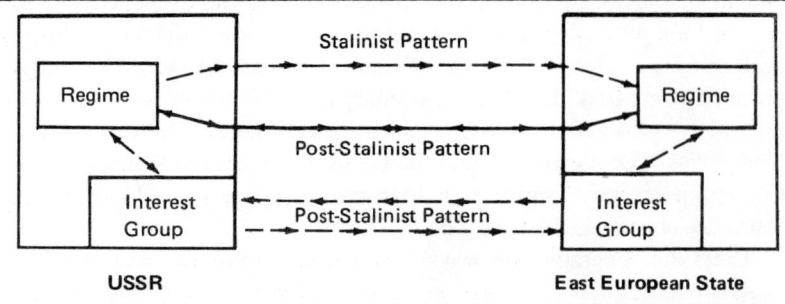

**Figure 1: DIFFERENCE BETWEEN LINKAGES IN THE STALIN PERIOD AND TODAY**

may have a decisive impact on internal developments. By and large the "reactive" linkages that exist do not take precisely the form that Rosenau outlines. The behavior of Soviet political actors is not simply "a response to behavior undertaken by" East European political actors, but also a *calculated use* of East European inputs, among others, for internal Soviet purposes. Of course, the invasion of Czechoslovakia was a "reactive" linkage in Rosenau's sense, but a more frequent "reactive" linkage occurs when a Soviet interest group or individual actor uses East European inputs as a weapon in the domestic political struggle. The difference between linkages in the Stalin period and today may be observed in the following simplified diagram (see Figure 1). In the Stalinist era linkages were between regimes, and innovation was spread from the Soviet Union to the East European states. At present, interest groups in East Europe may generate innovative proposals and communicate them to sympathetic interest groups in the USSR who might then convey the message to the Soviet regime, which, in turn, can send on its evaluation of the innovative proposal directly to the relevant East European regime. There is still a great deal of regime-to-regime communication, of course, but now the flow in innovative proposals moves in both directions. In both polities there is also a two-way flow between interest groups and the regime.

This pattern is made possible by the existence of what we called earlier a "concentric" or "dual" political system, which may again be represented by an overly simplified diagram (see Figure 2). A Soviet political actor who finds that there are obstacles preventing him from communicating and exercising influence in his "home" system may try to communicate through the now more highly integrated "associated" systems. Soviet Ukrainian writers, unable to publish in Kiev, found outlets in Prešov (Slovakia) and probably calculated that they would have a Soviet audience

Figure 2: CONCENTRIC OR DUAL POLITICAL SYSTEM

which reads Ukrainian publications or listens to the Ukrainian radio in Prešov. A Soviet political "liberal" might find it possible to publish in a Polish or Yugoslav journal, but not in a Soviet journal, and some of his Soviet supporters—and opponents—may read that Polish or Yugoslav publication. For example, the Soviet Jewish scholar, Yankl Kantor, published articles on Jewish writers and politicians who fell victims to Stalin's purges not in Soviet publications, but in the Warsaw Yiddish newspaper, Folksshtime. Presumably, the China-Albania outlet would be infrequently used, since only the most heterodox would find that channel appealing (the dissident Polish Communist Kazimierz Mijal, now residing in Tirana and broadcasting on Albania radio, is a quixotic illustration of the latter case). The Yugoslav outlet, too, is an uncertain one, and its use would depend on the standing of Yugoslavia in the alliance at any particular moment. The same would hold for nonruling, especially West European, Communist parties.

There are many forums and mechanisms through which the various forms of communication may flow. For regime-to-regime communications there are the usual diplomatic and Party channels, in addition to the Warsaw Pact and COMECON structures. For interest groups there are organizational delegations, tourism, contacts when in a third country at the same time, and simply reading each other's press and writing for it.

## RULES OF THE GAME

Having outlined some of the possible channels of communication of innovation, we turn now to a summary description of some of the rules of the game which seem to govern the process of transmitting an East

European innovation to Soviet decision makers, either in an effort to gain Soviet sanction to adopt and implement the innovation within the East European country, or, more ambitiously, in an attempt to get the Soviet Union itself to adopt the innovation. While Soviet adoption of an innovation will usually make that innovation available and legitimate for Eastern Europe, Soviet approval of an East European innovation less frequently results in Soviet adoption of that innovation. This seems to apply equally to ideas and institutions, though it is more difficult to trace the diffusion of ideas and measure their impact. Nevertheless, while the USSR has allowed East European countries to drop the insistence on socialist realism as the only permissible esthetic, it has clung to that doctrine within its own borders; while it has tolerated multiple-candidate elections in Poland and Hungary, it has not revised its own electoral practices accordingly. Thus East European innovations may be accorded legitimacy, at least tacitly, without themselves being adopted by the USSR. In fact, one would suspect that East Europeans are eager for the Soviet Union to adopt their innovations simply because of the legitimacy it confers upon their own ideas and actions, so that from the East European point of view it matters little whether an innovation is merely approved or actually adopted in the USSR.

Our problem is that there is no formal document or doctrine setting out the rules of the game, and we must derive them from the historical experience of the last fifteen years or so. The rules are not fixed and they are made, largely by the Soviet leadership, as situations arise. This makes the attempt to diffuse an innovation a hazardous one, since the penalties for rejection and refusal of legitimacy may be severe. Nevertheless, we can attempt to draw up some of the rules that seem to have emerged, especially in light of the experiences of Yugoslavia, Poland, Hungary, and Czechoslovakia.

(1) Perhaps the cardinal rule in innovation diffusion among socialist countries is that the innovator should never claim to be elaborating an alternative, competitive "model" of socialism. The Yugoslav "self-management system," the Chinese claims to have developed a model of socialism suited to "third world" countries and to have solved the problem of the transition from socialism to Communism, and the Czechoslovak claims, however muted and tentative, to the development of a form of socialism relevant to developed, industrialized countries, are examples of unacceptable claims which make it likely that rather than adopt the systems being developed the Soviet Union will deal with them with one or another of the rejection tactics. For reasons already discussed, it is very difficult for the Soviet leadership to accept explicit doctrinal claims which would call into

question its own doctrinal and, indeed, world-historical primacy. Even the Czechoslovak slogan of "socialism with a human face" probably aroused uneasiness in the USSR, because it implied that there might be inhuman socialism, and not in Czechoslovakia alone. The ideological imperative of presenting a unified front, the Soviet Union's self-image as the leader of the socialist world, defining the general nature of socialist programs and institutions, and the recent history of the splits in the Communist movement partially as a result of competing claims to leadership in innovation and development—all these make the Soviet Union wary and ab initio unreceptive to grandiose claims which pretend to the throne once occupied by the USSR alone.

(2) A related unacceptable element in an innovative program is the reduction of the role of the Party. As we shall see, this was the aspect of the Czechoslovak experiment that worried the Soviet leadership most, and for this reason the Hungarian regime is currently emphasizing that its economic reform will in no way diminish the role of the Communist Party in the political life of the country. The problem is who decides whether the role of the Party is being eroded or not. Unfortunately for Tito in 1948, and for Dubček twenty years later, that decision ultimately rests in the hands of the Soviet leadership. While Stalin's charges that the Yugoslav Communist Party was submerged in the national front and was a tool of the secret police—the irony of the latter claim was not lost on the Yugoslavs—may have been wholly specious, Soviet fears about the declining influence of the Czechoslovak Party seem more justified in view of the diminishing membership of the party, the revival of the Socialist and People's parties, the birth of new groups, such as Club 231 and KAN, and the fractionalization of the CPCS. The Party is the pivot of any Communist system, and any innovation which seriously affects its role and power is an innovation with serious implications for the very nature of the system.

(3) Any innovative proposal which would directly or indirectly involve withdrawal of the innovating Party from the Warsaw Pact and probably from COMECON would be unacceptable to the Soviet Union. The Czechoslovaks might have calculated that they have learned from the Hungarian experience in 1956 that withdrawal from the Pact was intolerable, and for this reason they tried to emphasize their fidelity to the Pact and its members. The problem for the Czechoslovak leadership was a dual one: to determine the 1968 rules of the game, and to maintain sufficient control of the situation and of the population so that they could play by these rules.

(4) Judging from the violent and sustained Soviet attacks on the literary and journalistic scene in Czechoslovakia in 1967-1968, we can deduce that the abolition of Party control of the mass media—i.e., censorship—is an intolerable step. There is a logic to this rule: the higher integration and level of interchange among socialist countries would make it difficult and costly to keep an uncensored socialist publication out of a socialist system where censorship was still maintained. As it is, relatively less censored socialist publications, such as the Polish *Polityka* and the Yugoslav *Politika*, have been read with eagerness by Soviet citizens.

(5) A more general and obvious prescription is that the proponents of innovation should seek to win the trust and confidence of highly placed Soviet politicians or interest group leaders. The innovators should also be careful in their choice of socialist allies, insofar as they have such a choice. The manifestations of support for the Dubček regime which came from Rumania and Yugoslavia in the summer of 1968 may have diminished, rather than enhanced, the legitimacy of the Czechoslovak innovators, for they raised the spectre of Czechoslovak heterodoxy in foreign and domestic affairs, as well as of a renewed "Little Entente," this time acting as a "non-conforming enclave" within the Communist world. Innovators must also try to cultivate the support, or at least the benevolent neutrality, of the relevant middle hierarchy in the USSR, and perhaps in other socialist countries as well.

(6) A corollary suggestion might be to avoid explicit, public approval in the West for the innovation or set of innovations. It is of course practically impossible to control the Western press from a socialist country, but there is no doubt that certain measures can be taken to reduce the chances that the innovative effort would be sensationalized in the West as a "loosening of the Soviet Bloc," "revolt against the Soviet master," "search for freedom," and the like. Judging by the great play the Soviet media gave to Western reactions to developments in Czechoslovakia from January to August, 1968, this is a sensitive point for Soviet leaders. Recent Hungarian attacks on the Western media for sensationalizing changes in Hungary indicate an awareness of this sensitivity.

(7) A further corollary is to avoid criticism, especially public criticism, of the USSR. One way of innovating in a resisting organization is to disguise the innovation. "Most organizations possess an underworld of technique and technology some of which is simply used to gain some freedom from the imposition of higher levels of authority, and some of which contributes to the achievement of corporate goals." [58] The innovation could be represented as something other than what it really is, something acceptable to the client. The Soviet leadership seems ever alert

to the danger of being fooled in this way, but the tactic might be used successfully, particularly if the innovation were not analogous to anything yet seen in the socialist repertoire.

(8) Brzezinski enumerates three stratagems for a deviant member of an alliance or organization:

- (a) the deviant acts as if there were no distinction between his and the center's doctrinal position. "This forces the center either to tolerate the deviation, allowing it perhaps to spread, or to take the first step in joining the issue, in effect launching a process of argument which it naturally would prefer to avoid. The skillful deviant makes the choice difficult by his strategy of ambiguity— always implying the possibility of his returning to the position of doctrinal subordination while gradually consolidating his position and probing the international movement for other sources of support."
- (b) The deviant can use the "asset of weakness" and plead inability to enforce orthodoxy (Poland's failure to collectivize was based largely on this tactic).
- (c) The deviant can employ the "asset of fanaticism... being more extreme than the center restrains the center by implying to it that the deviant 'has his heart in the right place but is a little extreme.'" [59]

It is worth repeating that the major difficulty with these—or any other—rules of the game is that they are not fixed and are decided pretty much ad hoc and unilaterally. It is as if one team in a basketball game decided as the game went along what constituted a legal score, a foul, or any other infraction. The opposing team could only hope to guess the next ruling or change of rules, and would be at a tremendous disadvantage in trying to develop a strategy or game plan. Until the ground rules for behavior in USSR/Eastern Europe are laid down and agreed upon by most of the protagonists this uncertainty will prevail.

## THE DIFFUSION OF
## POLITICAL INNOVATION: THREE CASE STUDIES

In this section we will examine some experiences, one still in progress, in the diffusion of political innovation from Eastern Europe to the USSR. It is from the Czechoslovak and Hungarian experiences in the recent years, as well as from the Soviet-Yugoslav rift of the 1950s, that we have derived our rules of the game, and by examining these cases we can observe the process in somewhat greater detail. In each case a brief profile of its defining characteristics will be drawn, and the Soviet reception accorded to the innovative proposals will be compared.

## Yugoslavia

The dispute between the Soviet Union and the Yugoslav Communist Party had its origins in World War II but emerged in a public debate beginning in 1948. The origins and causes of the dispute have been well documented and need not detain us.[60]

It must be remembered that, far from developing a new system or model of socialism in 1948, the Yugoslavs' initial reaction to Stalin's charges was to try to prove their superorthodoxy by slavishly copying the Soviet pattern in collectivizing rapidly and announcing a ridiculously ambitious industrialization plan. But the Yugoslavs were making the mistake of taking Stalin's charges too literally.

> The Russians were not interested primarily in what the rulers of Yugoslavia were doing with their absolute power, whether they were pursuing a pro- or anti-Great Serbia policy, whether the Communist Party of Yugoslavia was run according to its statute or not. By 1948 the Russians' main if not only interest was to replace Tito and the elite surrounding him by their own men. [The ideological and political charges against the Yugoslavs] came as a superstructure; they represented a rationalization, probably sincere, of the original and most important belief that Tito was a traitor since, more because of his position than by his choice, he could not be a pliable instrument in the hands of Moscow.[61]

Nevertheless, the charges against Tito are relevant to our inquiry because they reflect some of the "unacceptables" to Stalin. The method of dealing with Tito and his comrades, putting him beyond the pale of the socialist camp, should be noted as typical of the preferred Stalinist tactic for dealing with deviation.

Since in Stalin's time the doctrine of "multiple roads" to socialism had not yet been legitimated, anyone who objectively developed a different formula for the attainment of socialism, or was judged to have done so, was forced to develop an alternative model if he wanted to continue calling himself a socialist: the only way to reconcile self-identification as a socialist with identification by others as a renegade or a capitalist tool was to develop a different variety of socialism. In a way, Stalin forced the Yugoslavs to develop a competing socialist system. This the Yugoslavs did, beginning in 1950, with the "self-management system," including the workers' councils, the abandonment of strict controls in the arts, the abandonment of collectivization, the acceptance of aid from the West, the evolution of an independent foreign policy, and the promulgation of the eventual withering away of the Party, among other doctrinal and institutional innovations still in effect in Yugoslavia today.

Beginning in 1955, the Soviet leadership repudiated Stalin's treatment of the Yugoslavs and since that time there has been an uneven pattern of reconciliation and estrangement between the Soviets and the Yugoslavs. It would be futile to try to trace every twist and turn in Soviet-Yugoslav relations since 1955. We shall only note some points relevant to Soviet reactions to Yugoslav political-economic innovations within Yugoslavia itself.

In the Belgrade declaration of 1955 Khrushchev stated flatly that "questions of internal organization, or difference in social systems and . . . different forms of Socialist development are solely the concern of the individual countries."[62] This was a repudiation of the universality of the Soviet model and, if it were to be understood as a genuine statement of general principle, meant that there was some vaguely defined leeway for socialist systems to experiment and innovate in their political forms. It was a proclamation of *indifference* to Yugoslav innovation, something midway between rejection and adoption. Very soon after, however, perhaps under the pressure of the Hungarian and Polish events of the fall of 1956, *Pravda* attacked the Yugoslavs for proffering and promoting their alternative model of socialism to others.

> Is it right to denigrate the Socialist system of other countries, and to praise one's own experience, publicizing it as universal and the best? One cannot help but see that more and more frequently in the Yugoslav press the idea is appearing that the "Yugoslav road to socialism" is the most correct or even the only possible road for nearly all the countries of the world.

The same article commented favorably on the workers' councils but criticized measures which weakened central planning.[63] The Yugoslav "experience" was said to be irrelevant to the USSR on at least two counts: the USSR was much further along the road to Communism, and Yugoslavia's economy was so dependent on Western aid that its economic experience and system were irrelevant to all other socialist countries. Thus, the Soviets maintained a strategy of claiming a disinterest in Yugoslav innovations, based on their supposed irrelevance and unsuitability for adoption.

In 1958 *Kommunist* reacted to the recently announced Draft Program of the Yugoslav Communist Party by rejecting several of its major tenets: that capitalist states increasingly curb monopoly power, that is possible (in the Soviet version) "to arrive at socialism through a mere increased accumulation of Socialist features," that world tensions are caused not by the imperialists alone but by the existence of two blocs. The Yugoslavs were accused of ignoring the industrialization experience of the Soviet Union, of ignoring the "world historic fact" of the division of the world into two antagonistic systems, of talking prematurely of the withering of the state, and of emphasizing only the aspect of proletarian internationalism which guaranteed equality of Communist parties and nations, while neglecting the "necessity for strengthening unity and cooperation of the Socialist countries." Finally, the Soviets accused the Yugoslavs of "denying the leading role of the political party of the working class in a socialist state."[64]

Except for occasional sallies at Yugoslav economic failures as proof that their economic system as a whole, and their lack of collective farms in particular, were failed innovations, the Soviet leaders focused their criticism on Yugoslav doctrine in general and on Yugoslav positions in international relations, particularly as regards relations among Communist states. It was not Yugoslav "ideological deviation in internal policies" but "the interpretation of proletarian internationalism and the international political orientation of the Yugoslav state" that caused the Soviet Union to renew the dispute with Yugoslavia on a large scale in 1958.[65]

Soviet evaluation of Yugoslav systemic innovations, seemingly never detailed or systematic, fluctuates according to the prevailing state of Soviet-Yugoslav relations, and seems to be a relatively minor independent influence on those relations. There are two points that seem to be mentioned no matter what the tone of the overall treatment, namely, the diminution of the role of the Party and the "weakening of the planning element in the economy."[66] In 1967 and 1968, however, some dispassionate descriptions of the 1965 Yugoslav economic reform appeared in Soviet publications, and some articles took a mildly favorable tone.[67] By October 1968 the line seems to have shifted once more. An article in *Kommunist* returned to the familiar theme that Yugoslavia's economic difficulties, described in some detail, were due to the "eulogizing of the free market and a disparagement of the planning principle, which is under the socialist state's control." Equal emphasis was placed on "the erosion of the L.C.Y. [League of Communists of Yugoslavia]," with some lesser criticisms of Yugoslavia's attitude toward the Czechoslovak liberalization of 1968.[68]

In sum, the Soviet Union has criticized Yugoslavia mainly for its heterodox position on questions of international relations and the organization of the "socialist camp." It has not been particularly hostile—at least explicitly—to some of the innovative institutions of the Yugoslav self-management system. The Soviet Union has been consistently critical of certain doctrinal aspects of Yugoslav Communism, explicitly rejecting such innovations as the withering of the Party. It has also been consistently disapproving of the Yugoslav shift to a socialist market economy and of the diminished role of the Communist Party in Yugoslav political life and society. The USSR has expressed its disapproval explicitly and publicly, has not adopted any of the Yugoslav innovations to any significant extent, but has not dealt with the innovations as Stalin would have. That is, instead of trying to undo the innovations by force, or proclaiming that the innovations forced Yugoslavia outside the socialist camp irrevocably, the post-Stalin leadership has allowed itself the option of now embracing, now rejecting Yugoslavia, depending on particular considerations of the moment. Probably calculating that the danger of infection of other socialist countries by Yugoslav ideas and innovations is lessened, owing to declining Yugoslav prestige in Eastern Europe and the availability of the USSR to supply antidotes to such infections, the Soviet leadership does not feel impelled to launch a major onslaught on Yugoslav innovation. By harping on Yugoslav failures it calculates that it can neutralize the attractiveness of Titoist innovation.

**Czechoslovakia**

If in the Yugoslav case we can postulate that power considerations preceded ideological rationales for sanctions against a nonconforming part of the empire, in the Czechoslovak case we see ideological considerations as primary in motivating the USSR to impose power sanctions on a deviant member of the alliance.

The defining characteristics of the "Prague Spring" can be outlined as follows:

(1) The devolution of the power of the Communist Party, whether intended by the Party leadership or not, by permitting new political actors to appear on the scene (Club 231), or allowing old ones to assume new, more vital roles (trade unions, union of writers, youth organizations, Socialist and Peoples parties).

(2) The gradual erosion of Party control of the mass media, paralleling the yielding of Party monopoly of political activity and organization.

(3) Tentative feelers to like-minded deviant associates of the socialist camp (Yugoslavia, Rumania) and some overtures to the West, largely in the form of exploration of economic opportunities, particularly with regard to West Germany.
(4) Public criticism of the Soviet Union, both in the Stalinist period and as regards contemporary policy. There was much play given in the Czechoslovak media to the trials and purges of the 1950s and to the role of Soviet advisers in the terror period. There was also criticism aired of the position of the USSR regarding international relations among Communist states, and there were frequent criticisms of Czechoslovak—and, by implication or explicitly, Soviet—foreign policies, such as the stance on the Middle East.
(5) Some Czechoslovaks made claims that they were developing a model of socialism which could be profitably emulated by other developed countries, socialist and non-socialist. The official Party slogan was that Czechoslovakia was building "socialism with a human face."
(6) An extensive economic reform plan, originally articulated and approved under the Novotný regime, was adopted as an integral component of the new program.
(7) There was a thorough revision of nationality policy, including federalization of the country, benefiting the Slovaks, with ancillary benefits accruing to other nationalities, particularly Hungarians and Ukrainians in Slovakia.[69]
(8) In August 1968, new draft statutes for the Czechoslovak Communist Party were published. While maintaining the principle of "democratic centralism," the statutes contained some significantly new provisions: secret ballots for elections to Party organs, a time limit for holding offices, a curb on the power of professional Party employees, attempts to prevent the accumulation of great power in the hands of any one person, the federalization of the Party, allowing differing opinions to be voiced "provided these do not result in activity conflicting with the program and statutes of the party."[70]

Soviet response to these innovative actions and programs escalated from studied indifference, to interested neutrality, to criticism, to condemnation, and, finally, to repression by force. In January and February, 1968, the Soviet press paid scant attention to the changes in regime and subsequent developments in Prague, later increasing coverage of Czechoslovakia, but emphasizing international affairs and Soviet-Czechoslovak relations, rather than internal developments. Soviet criticism began to appear in May and June and escalated subsequently. Individuals, institutions, and officially approved innovations in ideology and political life were attacked. A survey of the major Soviet statements on developments in Czechoslovakia shows that the following themes were consistently emphasized: ideological revisions which downgrade the Leninist component of Marxism-Leninism; "discrediting" and "weakening" the Communist Party and its role in public life (abandonment of the principle of democratic centralism); the seizure of the mass media by "antisocialist and revisionist forces"; opposition to Soviet-Czechoslovak friendship; "statements that all of Czechoslovakia's 'woes' were related to the circumstance that until recently it was guided in its development by what someone called 'the Soviet model of socialism' "; "proposals for replacing principles of planning with unregulated market relations"; and, finally, revisions of Czechoslovak foreign policy in ways that would weaken the alliance of socialist states in Europe. [71]

The Czechoslovak case is a very good illustration of how particular East European political innovations, or, indeed, a political innovation of systemic proportions, become inputs into interest group conflict in the Soviet Union. On one level, Soviet difficulties with their own liberal intellectuals may have sharpened their interest in and fear of the prominence of liberal intellectuals as key political actors in Czechoslovakia. The trial of Galanskov, Dobrovolsky, Ginzburg, and Lashkova took place in January 1968. In May, *Pravda* criticized Soviet intellectuals who were not grateful enough to the CPSU and strongly implied that critical intellectuals were not to be legitimated as a political force.

On a more direct level, it has often been remarked that there were important linkages between liberalization in Czechoslovakia, intellectual unrest in the Ukraine, which was concerned not only with general political issues but with Ukrainian national grievances, and the intellectual community of the USSR as a whole.[72] Peter Potichnyj and Grey Hodnett have convincingly demonstrated that there was a difference of opinion in the Ukrainian political leadership wherein Piotr Shelest, First Secretary of the Ukrainian Party, spoke often and sharply against Ukrainian nationalistic manifestations, emphasized the great dangers raised by events in Czechoslovakia, and stressed the principle of central planning in economics, while V. V. Shcherbitskii, chairman of the Ukrainian Council of Ministers, seemed to lean toward experimentation with economic reform, displayed no undue anxiety over Ukrainian nationalism, and did not press the Czechoslovak issue.[73]

> As the Czechoslovak crisis unfolded, dogmatic forces in the Ukraine found it necessary and/or tactically desirable to expose to public view more directly ... the political meanings which were at stake in their struggle with the Ukrainian dissidents, and to hint that elements within the Party might have been finding these meanings

congenial to their own way of thinking.... Political attitudes were probably the aspect of Soviet system most significantly affected by the situation in Czechoslovakia.... The context here was the struggle between orthodox and liberal national-communist definitions of Ukrainian political realities; a struggle which had been going on for some years.... The importance of the battle of meanings was accentuated by the "ideological" character of Soviet politics and political culture, and by the assumption that "correct" doctrine has international validity. Although members of the liberal Soviet Ukrainian intelligentsia were undoubtedly the most avid consumers in the Ukraine of cultural and political wares produced in Czechoslovakia and Prešov [the Ukrainian center in Slovakia], it is unclear whether this group drew upon Czechoslovakia for theoretical concepts so much as for factual information and moral support. It is also difficult to distinguish the Czechoslovak influence from the more inclusive influence of liberal Marxist thought in Eastern Europe as a whole.... Given the "internationalist" (or imperial) outlook of Soviet officialdom, it is not difficult to imagine that it quickly projected onto the Czechoslovak scene fears and resentments which owed their origin to domestic ideological struggles in the Ukraine itself.

Czechoslovak "inputs," especially those originating in Prešov, added a valuable component to the already-existing non-official communications network in the Ukraine. Prešov played a substantial role in amplifying the dissemination of heterodox interpretations of political reality among the Ukrainian public ... introduced communicators into the network who were not inhibited by ordinary Soviet controls.[74]

There was a two-way flow in this linkage, with Soviet Ukrainians exploiting the "dual" political system by using the media in Prešov to transmit messages which could not be admitted by the Soviet media, "but which nevertheless could not expediently be labelled 'counterrevolutionary' when expressed through officially approved media in a fraternal 'socialist' society."[75] Thus, there was a linkage between Ukrainian interest groups in the USSR and Czechoslovakia, and through these linkage groups some of the outputs of the Czechoslovak innovations were transmitted to the Soviet Union where they became inputs into interest group and factional debates centering around domestic issues. These inputs could be used for contrary purposes. While a Shcherbitskii or a Ukrainian "nationalist" intellectual could choose to emulate or regard with benevolent interest the innovations in Czechoslovak nationality and political policy, a Shelest and other Soviet officials could point to them as object lessons in the dangers of such innovations.[76] Czechoslovak influences had been superimposed upon processes indigenous to the Ukraine, processes visible as early as the first years of the 1960s. For Soviet leaders of "hawkish" inclination, the Czechoslovak "threat" to the Ukraine might well have provided *one* extremely convenient peg on which to hang their case. It vividly dramatized the danger of developments in Czechoslovakia, and did so with reference to the "Ukrainian question"—a problem about which many members of the Central Committee were more likely to have been anxious than well-informed.[77]

The case just discussed sharpens and clarifies some of the contemporary rules of the game of the diffusion of political innovation from an East European state to the Soviet polity, and also illustrates the way in which externally generated political innovations can become inputs into political debate in the Soviet Union.

## Hungary

Our third case study is very much in progress—the elaboration and implementation of economic reform and political adjustments in Hungary in the late 1960s. This is a further test of the rules of the game and a great deal of attention has been paid to them by the Hungarian innovators. This case will again illustrate the use of East European innovations in Soviet politics. The Hungarian innovators of the 1960s seem to have profited greatly from two historical lessons—their own searing experience of unsuccessful deviation in 1956, and the more recent, and perhaps equally instructive, fate of the Czechoslovak innovators.

Some important differences between the ways in which the Czechoslovak and Hungarian innovations have been formulated ought to be pointed out before presenting a profile of the Hungarian innovations.

While both the "Prague Spring" and the less dramatic reforms in Hungary were (are) attempts to reestablish the legitimacy of a political regime, and even of a system, which had suffered a severe loss of authority, the Czechoslovaks tried to reintegrate state and society largely on a normative basis, whereas the Hungarians have stressed the use of material incentives for reintegration. The Czechoslovak reformers appealed to the population on the basis of a new political program containing modified values, whereas the Hungarians have deemphasized value and ideological elements in their attempt to offer better material prospects as a way of reconciling the Hungarian population to the system. Since the Czechoslovak reintegration was attempted on a normative basis, it was forced into explicit ideological revisions, evoking a great deal of Soviet concern and anxiety. Furthermore, since the Soviet leadership does not perceive the need for ideological revision of its own system, it was difficult for it to empathize or even sympathize with its counterparts in Prague. The Hungarians, however, by focusing on material incentives and changes in the production and distribution of those incentives, have not been forced to confront ideological issues or articulate them in as direct a way as the Czechoslovaks had been. Moreover, since the USSR itself has perceived the need to tinker with the economic structure and reward system of its own socialist economy, its leadership can better understand Hungarian motivations and actions.

Secondly, while the Czechoslovaks were forced by their normative concerns to present their innovations as innovations indeed, the Hungarians are better able to present the changes in their system as mere tinkering with a fundamentally unchanged system, one which basically remains a replication of the Soviet model. The Hungarians see little if any advantage in presenting their innovations as having great systemic implications, and they have tried to avoid discussing their innovations in abstract, ideological terms. This may be one reason why Andras Hegedus, who tried to initiate a discussion on the "alternatives of social development," was criticized and demoted, the proposed discussion never really taking place in the way Hegedus seems to have urged.[78]

A third important difference between the two cases is that the Czechoslovak innovations were evolved in the heat of political battle by a new, unstable political leadership not very well known to the Soviet leadership. The Hungarian innovations were mapped out gradually, in relative calm, by a leadership which was stable, apparently united, and, from a Soviet perspective, reliable and trustworthy. Because of leadership stability, Hungary could "go slower," while in Czechoslovakia both the population and rapidly evolving events themselves pressured the leadership for immediate change. In Hungary the existing machinery was able to contain and process the innovative proposals, while in Czechoslovakia the machinery itself was in the process of transformation and was being fought over. Finally, just as the Poles had learned the limits of change from the Hungarian experience of 1956, the Hungarians learned from the Czechoslovak experience in 1968.

Harry Shaffer points to another difference which seems significant. In the 1960s Hungary's economic progress was neither as rapid as that of Bulgaria or Rumania, nor as weak as that of Czechoslovakia and Yugoslavia

> where corrective steps became a matter of utmost urgency. Hence, while radical changes in Hungary's economic system were necessary if economic performance was not to deteriorate, the house was not on fire and there was time for thorough preparation for whatever alterations were to be introduced.[79]

Our profile of the Hungarian innovations will be somewhat more detailed than the Czechoslovak one since they are less well known and less dramatic.

(1) The Hungarians constantly stress that the leading role of the Party is in no way diminished by the economic reforms which went into effect January 1, 1968. One Hungarian spokesman clearly implied that the Hungarians would not fall into the Czechoslovak heresy on the question of the Party's role.[80] A Soviet author, writing in the authoritative Party journal *Kommunist,* noted that in the period of economic reform Party ideological work was all the more crucial and that the Hungarian Party

> appears as the leading and directing force in the political union of Communists and non party people.... The growth in the leading role of the HSWP [Hungarian Socialist Workers Party] is inseparably linked with the further perfection of the organizational structures of the Party and the development of intra-Party democracy.[81]

"The economic reform embraces our entire society and is of great political significance because the leading force of society—the Party—must manage its determination," said a Hungarian economist writing in a Soviet economics journal.[82] Of course, rhetoric and reality may not coincide, and it remains to be seen whether the Hungarian Party, in a "low profile" position ever since 1956, will maintain or increase its role in the political, social, economic, and cultural spheres.

(2) The Hungarian economic reform itself is perhaps the basic element in the more general reform. "While the fundamental features of the Hungarian reform may be similar to those of the reforms already introduced in other Communist countries, the *extent* to which the reform has been carried is more far-reaching in Hungary than anywhere else in Eastern Europe except Yugoslavia."[83] Without detailing the major features of the reform, we can mention the introduction of a flexible "three category" price system, the power given to large enterprises to make their own export contracts with foreign enterprises, the application of reform principles in agriculture as well as industry, and a degree of independence for enterprise directors which "appeared to be greater than elsewhere in the bloc. Finally, the Hungarians were the first to realize that in the changed conditions of the new economic model, the role of the trade unions would have to undergo significant reassessment." One of the essential characteristics of the Hungarian reform is that

> The central [planning] organ is to fix only the long-term tasks to be fulfilled, but the details of these tasks will be set by the enterprises

themselves according to the market requirements. It will be the duty of the superior organs to harmonize these market requirements and national interests.[84]

By and large, the Hungarians seem to have deliberately avoided discussing the economic reform in the wider context of systemic crisis and the need for systemic change. That is, they have avoided the kind of rhetoric that characterized the work of Ota Šik, for example, perhaps simply because the Hungarian reform was not an emergency response to a deeply felt crisis, but a measure taken rather deliberately, as something which ought to be done sooner rather than later to avoid the kind of system breakdown confronting the Czechoslovaks. This enabled the Hungarians to present their reform more as a technical adjustment in the system than as a major overhaul and reevaluation of its basic machinery and workings. "Instead of chasing 'hazy pictures' or the model of an 'abstract' democracy which can be realized nowhere, the Hungarian regime seeks solutions corresponding to domestic realities."[85] Jozsef Bognar put it clearly:

> It is not expedient to include in the debate and to expose to social confrontation the practical and theoretical multiplying effects of the changes when the reforming movement is launched, for in this case the battle-lines become confused and this, as a rule, encourages the opponents of the change.... It is better to deal with the secondary and accessory consequences when agreement has been reached on basic issues.[86]

(3) Part of the economic reform has been an increase in the power of the trade unions. The New Labor Code gives workers the right to share in enterprise profits and makes the unions responsible for the protection of the material interests of the workers, by giving them the right to a decisive voice in determining the percentage of profit which is to go into the profit-sharing fund. The unions also share with management the duty of drawing up the annual collective contract which outlines the fundamental duties and rights of all parties in the enterprises. This is based on the recognition that managerial and production personnel may have interests which are not congruent. The most novel labor right included in the New Economic Mechanism is the right to exercise a veto in three kinds of situations: when enterprise actions violate the terms of the collective contracts, when working conditions fall below minimum safety standards, and when the enterprise takes steps which violate "socialist morality." The use of the veto can halt implementation of managerial action until the dispute is settled by a governmental body. In a dispute in the Athenaeum

publishing house, the union vetoed a managerial plan to impose overtime work whereby the funds for the overtime would come from the profit-sharing fund. When the dispute was arbitrated the union won a complete victory.[87]

(4) It is important to note that changes in the status of the labor force took place by altering existing institutions, not by creating new ones. For example, the Hungarians might have created workers' councils of some sort to replace or supplement the trade unions, but they refrained from doing so, probably because of their own unhappy experience with such councils in 1956, and because this would unnecessarily dramatize the nature of the changes taking place, perhaps rousing Soviet fears that a Yugoslav model was being experimented with.

(5) The Hungarian government has extended the economic reform to the agricultural sector. By 1966 about two-thirds of all collective farms had adopted the system, first introduced on the Nadudvar collective, whereby each family or other field unit is assigned a particular parcel of the land and the collective farmer gets a specific percentage of the total crop produced on that parcel, no matter how high the yield. "This payment is in addition to the income from the work units accomplished [trudoden'] and is paid for in cash."[88] This guarantees the farmer a minimum income no matter the size of the yield and helps improve and stabilize peasant income. "It is perfectly clear that a most important and fundamental tenet of collectivization has been repudiated, if not entirely abrogated, by Nadudvar: collective membership working collectivized land for the common (collective and state) good."[89] As shall be seen, this innovation became an input into a Soviet debate.

(6) Just as in the labor sector familiar institutions have been given new rights and powers, so in the political sphere instead of elaborating new institutions and processes the Hungarians have given parliament new importance, strengthened the National Front, and changed the electoral system. Beginning in 1967, or perhaps a year or two earlier, a new legislative "style" was inaugurated,

> thanks to which the members of the House no longer simply approved of the bills worked out by the government, but actually "made" the laws themselves by participating in the elaboration of the drafts. . . . The new, correct, and now established practice is to place the drafts of bills prepared by the competent ministers before the pertinent standing committee in order to enable its members to suggest modifications. Subsequently, the ministries draw up the final draft, which comes before the Council of Ministers for approval with a view to the changes recommended by the standing committees. This means that the bill submitted by the Council of Ministers to the

Plenum of the House already "incorporates" the views of the deputies sitting in the various standing committees [sic]. [Speaker of the House Gyula] Kallai disclosed that the standing committees of Parliament had held 34 meetings and that there were "brisk" debates over the drafts of bills.[90]

In an interview in *Izvestiia*, Kallai stated that "in spite of the fantasies coming out of some of the capitalist countries, reality confirms that the parliament plays an indispensable role, with great significance in a socialist society in general, and especially in the current stage of development of our country." He pointed out that ministers have been challenged on the floor of Parliament and that

> the work of parliament was noticeably animated, discussions were set off more often, creating a healthy critical spirit. The minister-speakers on this or that question now not only "give a speech" but also lend an ear to the discussions.... We are trying to develop the initiative of the deputies ... to encourage the deputies to join in as initiators of discussions in parliament of important government problems.

Kallai also stressed the importance of the right of interpellation by a deputy.[91]

(7) The Hungarian regime has conducted a limited experiment in allowing several candidates to stand for a single office, and it has changed the electoral system so that each district elects a deputy. Kallai claimed that allowing two or more candidates to stand for Parliament resulted in a greater feeling of responsibility toward the constituency on the part of the deputies and a greater familiarity with constituency sentiment.[92] In the April 1971 parliamentary elections, 49 of 352 parliamentary seats were contested, and in local elections over 3,000 of 70,000 offices were contested.[93]

(8) The general tenor of the Hungarian reforms is that of reconciliation with an alienated population. Kadar's famous 1961 formulation of "He who is not against us is with us" is a succinct expression of the Hungarian attitude toward the population and the willingness of the regime to resocialize the population gradually, settling for a slow increase in the political consciousness and loyalty of the population. This is accompanied by a "struggle on two fronts," an attempt to walk a middle path between the Scylla of "dogmatism" and the Charybdis of "revisionism." The Hungarians are attempting to avoid the excesses both of the Rakosi regime and of the short-lived Nagy period, and to choose a middle way between, say, a Rumanian or Bulgarian posture, and a Czechoslovak (1968) experiment.

The preference of the Hungarians for concentrating on "small deeds" rather than "senseless dreams," to borrow terms from Russian history, is reflected, for example, in the quiet filling of every one of the many vacancies that have existed in the Catholic church hierarchy for about eighteen years.

> This quiet mutual acceptance of an uneasy status quo attempted to normalize the internal situation and thus placed the entire Mindszenthy case ... in the perspective of a meaningless anachronism. This across-the-board agreement between the Catholic clergy and communist authorities ... attests to the internal stability and surprising consolidation of the Janos Kadar regime.[94]

Another symptom of the regime's posture is the truly astounding fact, especially in the East European context, that "For a decade the regime and the literati have been living together in peace, avoiding any dramatic clashes. This peaceful coexistence was preserved even at the height of the Czechoslovak crisis."[95]

Although Hungarian officials would probably object to the use of these terms, there has been a gradual depoliticization of some areas of Hungarian life, literature being only one instance. A crucial step was the deemphasis on political criteria for managerial personnel in the economy and an increased emphasis on professional criteria. "The campaign against party hacks began early in the 1960's when the Kadar regime started appointing non-Party experts to important administrative posts."[96] Class and other social criteria were dropped in considering applications for admission to institutions of higher education. The very definition of "proletarian" was broadened so that people who were not workers at the bench could enjoy the rights and privileges of the working class. "In other words, the insuperable cleft between a significant portion of worker and intellectual occupations has ceased to exist, and this characteristic feature—one could say qualitative change—of the workers' class is valid many times over the respect to Communist workers."[97] According to the editor of Népszabadsag, the alliance of the Party with non-Communists is a necessary precondition "of successful Communist construction. ... We regard the further development of socialist democracy to lie in the strengthening of this alliance."[98]

In sum,

> We have carried out this social conciliation in accordance with the changes in social and political conditions, neither sooner nor later than the given situation required because both possibilities would

have caused damage. And even in recent years we have been involved in constant debate both with those who unreasonably and rashly tried to push us forward and with those who wanted to push us back to a conservative standpoint which abhors democracy.[99]

(9) Throughout the period of economic and political changes, the Hungarians have stressed that the alliance with the Soviet Union is not diminished but strengthened, and that Hungary is very much aware of its responsibilities to the socialist community. The Prime Minister emphasized that in conceiving the NEM the regime was careful to ensure that "international public opinion," and especially the "socialist community," would understand what was being done and, hopefully, would agree with it. "Our place is at the side of the Soviet Union, which is the fundamental force in the socialist concentration, in the struggle against imperialism and in the reshaping of internationalist unity."[100] Soviet commentators, as well as Brezhnev himself, have frequently noted with approval the emphasis that the Hungarians place on their alliance with the USSR.

> The Party attaches great importance in its propaganda to the idea of proletarian internationalism and socialist patriotism, to the further development and consolidation of the friendship between the Hungarian people and the people of the Soviet Union and other fraternal countries, to the strengthening of the unity of socialist cooperation and the world communist movement. The Party consistently and decisively denounces chauvinist anti-Soviet "national communism."[101]

Obviously, invidious comparisons with wayward Czechoslovakia are implied, especially in the Soviet sources.

At the Tenth Congress of the Hungarian Socialist Workers' Party in November 1970, Brezhnev praised Kadar as "the true son of the Hungarian people, an outstanding and respected figure in the international Communist and workers' movement." He went on to praise the economic reform, the policy of "struggle on two fronts," and, in much vaguer terms, the political reforms ("expansion of socialist democracy").[102]

(10) Closely connected with the emphasis on ties to the socialist community, and to the Soviet Union in particular, is the theme of Western misinterpretation and misuse of the developments in Hungary. This is in keeping with our rule of not being made to look in the Western media like a deviant from Soviet norms. Imre Pardi wrote that

> in the West, beginning with the political aspirations and calculations of reactionary circles, the reform is depicted as a sort of liberalization process in social life resulting in a weakening of socialist social

relationships . . . an estrangement from the Soviet Union and a weakening of socialist international cooperation. I consider that there is no need to waste words on the fabrications spread by Western bourgeois circles. [103]

*Izvestiia's* correspondent in Budapest wrote an article, significantly entitled "Dynamism without Sensationalism," in which he pointed to Western attempts to exploit changes in Hungary for anti-socialist and anti-Soviet purposes. A Hungarian journalist devoted an entire article in *Izvestiia* to this subject, [104] and the *Izvestiia* correspondent wrote a heavily sarcastic article about a Zurich newspaper's depiction of Hungary as "disillusioned and insulted." [105] Several articles in this vein have appeared in the Hungarian press as well. [106]

(11) Connected with this effort to play down the liberalizing nature of Hungarian innovation, is the insistence that Hungary is not elaborating a competitive or even distinctive "model." Prime Minister Fock argued that the Hungarians had rejected "provincial" interpretations of socialism and that NEM is not a Hungarian invention since similar problems and solutions had been discussed in other socialist countries. [107] Hungarian commentators have repeatedly pointed out that, on the one hand, their experience proves that there is no one single model which all Communist countries must follow, but, on the other hand, Hungarian variations on socialist themes do not constitute a comprehensive and distinct model, and certainly not one which can be presented as an alternative to the mythical Soviet model. Thus, the Hungarians seem to have been very careful to observe our cardinal rule of the game.

## HUNGARIAN INNOVATIONS AND SOVIET DOMESTIC POLITICS

We conclude with an examination of Hungarian innovations as inputs into Soviet domestic issues. First, we must note that the generally favorable treatment the USSR has accorded the Hungarian innovations, particularly the NEM, is due to the nature of the innovations themselves and Hungary's ability to abide by the rules of the game. It is far too simplistic to say, as William Robinson does, that

> the Soviet attitude has tended to confirm that in many cases it is really not the nature of the ideological "heresy" which arouses the condemnation of the CPSU, but the trustworthiness of the "heretic" instead. That is to say, the Soviet Union has confidence in Kadar's ability to handle reform and trusts him to safeguard bloc (i.e., Soviet) interests. Dubček, however, was never held in such "high esteem." [p. 4].

This is only a small part of the explanation, as we have tried to demonstrate.

When we speak of the Soviet reaction to Hungarian innovation, we can pretty much exclude adoption by the Soviet Union of the entire package of Hungarian institutional innovations, simply because both parties agree that some Hungarian experiences are irrelevant to the Soviet Union, especially in view of the greatly different sizes of the two countries.[108] We hypothesize, however, that Soviet political actors who favor further economic reform in the USSR, whether along Hungarian lines or not, have chosen the Hungarian reform as a surrogate for arguments about Soviet reforms and have praised NEM as a way of pointing to the potential advantages of a Soviet NEM. By the same token, Soviet opponents of economic reform raise questions about the NEM as indirect opposition to reforms at home. Similarly, some Soviet citizens have commented favorably on political changes in Hungary in order to recommend similar changes at home.

It may be assumed that those Soviet correspondents and commentators who are reporting favorably about innovations in the Hungarian system are either themselves enthusiastic about them or are following the wishes of higher-placed people—editors and politicians—who are anxious to show the Hungarian reforms in a favorable light. Correspondent Rodinov went so far as to say that Hungary was "taking the lead, in fraternal union with the USSR" in socialist construction. The same correspondent praised Hungarian reforms in agriculture and pointed to the powers enjoyed by the Hungarian National Council of Cooperatives, and used to protect Hungarian farmers, while no such protective devices exist in the Soviet Union. Since the article appeared at a time when the Soviet authorities were working on a revised charter for Soviet collectives, it may well have been directing attention to Hungarian institutions as a model for emulation.

> The ground rules for a decompression of a centralized agricultural system ... have been staked out in the *Izvestiia* account. ... The zest with which *Izvestiia* covers the Hungarian experience nourishes the hope that ultimately a similar development of Kolkhoz democracy and market order will be allowed to flower in the Soviet Union.[109]

There seems to have been a clear-cut difference of opinion in the Soviet agricultural establishment on the Nadudvar system. The secretary for agriculture of the Soviet Party's Central Committee inspected the Nadudvar collective in 1964, the chairman of the collective visited the

USSR, and an article on the Nadudvar system appeared in *Kommunist* (written by a Hungarian) in the same year. "Publication of this article was an important sign of Soviet willingness to discuss, and even reconsider, the uncompromising Soviet system of collectivization; it also indicated approval of Hungarian agricultural policy."[110] The USSR then experimented with the Nadudvar system in a few of its own collectives and in some ways seems to have carried the experiment even further than the Hungarians. While this was going on, the generally conservative agricultural newspaper *Sel'skaia Zhizn'* maintained complete silence on both the Hungarian and the Soviet experiments, apparently a sign of opposition to the innovation. In March 1967, a *Pravda* editorial came out against changes in central planning and increased incentives in agriculture, but in August 1968, *Pravda's* correspondent in Hungary published a laudatory article on Nadudvar.[111] All this seems to add up to continued debate within the Soviet agricultural hierarchy on the merits of various systems of agricultural organization and incentives. In this debate the Hungarian innovation at Nadudvar, clearly an innovation with systemic implications, serves both as a model and as a symbol.

Similar Soviet interest group activity can be observed around the Hungarian NEM in general. Aside from the laudatory treatments of the NEM, there have appeared in the Soviet press some carefully worded critical commentaries. Professor Rem Belousov, an official of the Soviet Planning Commission (Gosplan), said in an interview in a Western newspaper, "We are attentively watching how it [Hungarian NEM] turns out. Should it fail, then the Soviet Union can help Hungary. We ourselves, however, must be very careful with such far-reaching experiments."[112] Belousov was expressing mild skepticism about NEM itself, but he was also expressing more clearly his doubts as to its relevance to the Soviet Union. He may have had a valid point when saying that the Soviet Union could bail Hungary out in case of the NEM's failure, but who could bail the USSR out in case its enormous economy would suffer reverses as a result of economic reform? Not Hungary, certainly. Therefore, the USSR "must be very careful" with such ventures.

A second Gosplan official told a Western economist that the Hungarian reform is very similar to the Czechoslovak, as indeed it is. "We do not like to criticize our Hungarian comrades, but we think their approach to central planning is totally incorrect; and we should know, because we here in the Soviet Union have had 50 years of experience with central planning."[113] Two things are worth pointing out: a Soviet official claims that by virtue of its historical position the USSR can better evaluate proposed innovations than can its junior partners; Soviet experience with central planning is much greater than that of Hungary's and therefore, perhaps, the Soviet experience ought to be emulated. It is very important to note that the two Soviet economists we have cited as somewhat opposed to NEM and its adoption by the USSR are associated with Gosplan, the main agency involved in central planning. Their institution, and their own personal positions, would be diminished in importance and power by the adoption of an NEM type reform which deemphasizes central planning. These Soviet officials have personal stakes in opposing innovation of the NEM type. A recent report from Moscow also emphasizes that Gosplan has been opposed to the kind of decentralization involved in economic reforms Hungarian style. "At a recent press conference, A. V. Bachurin the agency's deputy director said that recent events 'graphically have shown where the illusions spread by some theoreticians can lead'" and he quoted Lenin as supporting central planning, promising that the next Soviet five year plan will "give priority to tighter planning and more centralization, not less as reformers originally envisaged when Premier Kosygin made the reform public in 1965." [114]

Two other examples of the use of external inputs in internal debate may be cited. One is a statement appearing in *Novyi Mir,* generally considered to be a medium of expression for Soviet liberals. Reviewing several Soviet publications, V. Savin made so bold as to mention political innovations which he suggested might be tried in the USSR. For example, he quoted Lenin—always a good political tactic—to the effect that opinion groups ought to be allowed to publish petitions and the like in the public press. Considering the nature and source of some recent Soviet petitions, this was a bold proposal indeed. Savin also mentioned that "several scholars ... have suggested the necessity of increasing the importance of voting in the elections to the Soviets. This suggestion, it seems to us, deserves discussion. For example, in socialist Hungary recently the following electoral system was introduced," and he went on to describe the reforms discussed earlier, adding that "a similar system exists in some other socialist countries." [115] The tactics employed by Savin are instructive: Yugoslavia has developed the electoral system to a much greater extent than Hungary, but Yugoslavia is not as "respectable" in the USSR; Poland also has multicandidate elections, but the Polish regime is certainly no symbol of liberalization, nor, on the other hand, has it gotten as good a press recently in the Soviet Union as has Hungary. Finally, Hungarian *agitprop* methods were favorably commented upon by V. I. Stepakov, then head of the CPSU Central Committee's *Agitprop* Department, following a tour of Hungary. According to Aryeh Unger, Stepakov was very much involved in a debate as to whether the traditional Soviet agitator was to be replaced by a "politinformator," and he used the purported Hungarian success in *agitprop* as evidence of the soundness of his own position.[116] Once again, we observe the use of a foreign innovation as a surrogate for Soviet issues and, more than that, as a model explicitly held up for emulation.

## INNOVATION AND POLITICAL CHANGE IN COMMUNIST EAST EUROPE

The history of the diffusion of political innovation from Eastern Europe to the USSR is instructive in pointing out certain aspects of the Soviet-East European relationship and in highlighting some of the goals and priorities both of the Soviet and of the East European leaderships. The tortuous and uncertain course of innovation diffusion vividly illustrates the changing and highly unstable relationship of the USSR to Eastern Europe in the post-Stalin era. It points to the likelihood that the East European states will continue to be caught in the dilemma of having to innovate in order to strengthen their own internal legitimacy, efficiency, and stability, but at the same time having to satisfy their external audience as well. The partial autonomy gained by the East European states in the last fifteen years has made their political task more difficult, for they are now expected to achieve domestic successes on their own and without the weapons of coercion, while at the same time the ways in which they can move toward such successes are severely circumscribed by the USSR.

Secondly, attempts at innovation diffusion in an era of polycentrism and relative institutional experimentation dramatize and bring into sharper relief the prevailing institutional and ideological orthodoxies obtaining in the USSR, even though it appears that those orthodoxies are quite often latent, activated only by challenges from within and without. The uncertainties of Soviet orthodoxy and policy are what make the innovative process so risky, and the process itself helps to establish in practice the nature and degree of change in Soviet theory and practice.

Finally, the recent experiences of the East European innovators yield clues as to the short-range possibilities and probabilities of internal change in Eastern Europe as well as in the Soviet-East European relationship. The history of the Czechoslovak and Hungarian innovations and their treatment by the USSR point out the severely constraining boundaries within which change in the socialist systems of Europe is possible.

## NOTES

1. Anatole Shub, "Lessons of Czechoslovakia," Foreign Affairs, Vol. 47, No. 2 (January, 1969) p. 273.
2. Richard V. Burks, Technological Innovation and Political Change in Communist Eastern Europe, RAND Memorandum RM-6051-PR. Santa Monica, California: August, 1969, p. 59.
3. Zbigniew K. Brzezinski, "The Organization of the Communist Camp," World Politics, Vol. XIII, No. 2 (January, 1961), p. 204.
4. Ghita Ionescu, The Break-up of the Soviet Empire in Eastern Europe (London, 1965), p. 7-8.
5. V. Kotyk, "Some Aspects of the History of Relations among Socialist Countries," Československy Časopis Historicky, No. 4, 1967. Radio Free Europe, Czechoslovak Press Survey No. 1973, p. 14. For an interesting and orginal treatment of relations among socialist countries in the Stalin era, see Kotyk's Svetová Socialistická Soustava (Praha, 1967), pp. 5-103.
6. Ole Holsti and John D. Sullivan, "National-International Linkages: France and China as Nonconforming Alliance Members," in James N. Rosenau ed., Linkage Politics: Essays on the Convergence of National and International Systems (New York, 1969), p. 164.
7. R. V. Burks, "The Communist Polities of Eastern Europe," in Rosenau, op.cit. The classic work on the evolution of Communist East Europe is Zbigniew K. Brzezinski, The Soviet Bloc, (Cambridge, Mass. 1961 and 1967).
8. Irving Louis Horowitz, "Consensus, Conflict and Cooperation: A Sociological Inventory," Social Forces, Vol. 41, No. 2 (December, 1962) p. 187. Jowitt applies Horowitz's distinction in "The Romanian Communist Party and the World Socialist System: A Re-definition of Unity," World Politics Vol. XXIII, No. 1 (October, 1970).
9. For a useful summary of debates on "different roads to socialism," see Paul Kecskemeti, "Diversity and Uniformity in Communist Bloc Politics," World Politics, Vol. XIII, No. 2 (January, 1961).
10. See Fundamentals of Marxism-Leninism (Moscow, 1963), esp. Ch. 25, and Sh. P. Sanakoev, Mirovaia Sistema Sotsializma (Moscow, 1968), esp. part two.
11. Nish Jamgotch, jr., Soviet-East European Dialogue: International Relations of a New Type? (Stanford, 1968) p. 127.
12. Jan Triska, "The World Communist System," in Triska, ed., Communist Party-States (New York and Indianapolis, 1969), pp. 33.
13. Lawrence B. Mohr, "Determinants of Innovation in Organizations," American Political Science Review, Vol. LXII, No. 11 (March, 1969) p. 112.
14. Karl W. Deutsch, The Nerves of Government (New York, 1966) pp. 207 and 221.
15. Ibid., pp. 247-248.
16. Jack L. Walker, "The Diffusion of Innovations Among the American States," American Political Science Review, Vol. LXIII, No. 3 (September, 1969), p. 889.
17. H. G. Barnett, Innovation: The Basis of Cultural Change (New York: 1953), p. 375.

18. Ibid., p. 376.
19. Everett M. Rogers, Diffusion of Innovations (New York, 1962) pp. 124-131. On the importance of compatibility with previous value patterns, see also Deutsch, p. 148.
20. Ibid., p. 131.
21. Walker, p. 897.
22. Deutsch, p. 147.
23. James Q. Wilson, "Innovation and Organization: Notes Toward a Theory," in James D. Thompson, ed., Approaches to Organizational Design (Pittsburgh, 1966), p. 208.
24. Ibid., p. 210.
25. Rogers, p. 99.
26. Karl W. Deutsch, "External Influences on the Internal Behavior of States," in R. Barry Farrell, ed., Approaches to Comparative and International Politics (Evanston, Ill., 1966) p. 11.
27. Ibid., p. 24.
28. Warren G. Bennis, "Theory and Method in Applying Behavioral Science to Planned Organizational Change," in Bennis, Warren G., Kenneth D. Benne, and Robert Chin, eds., The Planning of Change (New York, 1969) p. 77.
29. Ibid.,
30. Barnett, 93.
31. Bennis, 76.
32. A. L. Kroeber, "Diffusionism," from an article by that title in Edwin R. A. Seligman and Alvin Johnson, eds., The Encyclopedia of the Social Sciences, III (New York, 1937). Excerpted in Amitai and Eve Etzioni, eds., Social Change (New York, 1964), p. 143.
33. Goodwin Watson lists five sources of resistance to change in social systems: (1) conformity to norms or habits; (2) systemic and cultural coherence; (3) vested interests; (4) "the sacrosanct" ("The greatest resistance concerns matters which are connected with what is held to be sacred."); (5) rejection of "outsiders." We mean by "defining characteristics" something like Watson's "Sacrosanct." See Watson, "Resistance to Change," in Bennis, Benne, and Chin, op. cit. For another listing of factors inhibiting innovation acceptance, see Ronald G. Havelock, et al., Planning for Innovation Through Dissemination and Utilization of Knowledge (Ann Arbor, Institute for Social Research, 1969) pp. 6-7–6-10.
34. Deutsch, "External Influences," p. 11.
35. Rogers, p. 280.
36. Herbert A. Shepard, "Innovation-Resisting and Innovation-Producing Organizations," in Bennes, Benne, and Chin, p. 520.
37. Ruth Leeds, "The Absorption of Protest: A Working Paper," in Bennis, Benne, and Chin, p. 199.
38. On the notion of transitive and reflexive goals, see Lawrence B. Mohr, "The Concept of Organizational Goal," Institute of Public Policy Studies Discussion Paper No. 9, University of Michigan, September 1969.
39. On "trained incapacity," see Robert K. Merton, Social Theory and Social Structure, (Glencoe, Ill., 1949) pp. 153-156.
40. Leeds, p. 201-202.
41. Ibid., p. 204.
42. Ibid, p. 207.

43. Ibid., p. 208.
44. Rolf P. Lynton, "Linking an Innovative Subsystem into the System," Administrative Science Quarterly, Vol. 14, No. 3 (September, 1969), p. 400.
45. Burks, Technological Innovation, p. 33.
46. See the writer's "Power and Authority in Eastern Europe," in Chalmers Johnson, ed., Change in Communist Systems (Stanford, 1970).
47. Holsti and Sullivan, p. 164-65.
48. Zbigniew Brzezinski, "Deviation Control: A Study in the Dynamics of Doctrinal Conflict," American Political Science Review, Vol. LVI, No. 1, (March, 1962) p. 6.
49. Jamgotch, pp. 105 and 108.
50. Holsti and Sullivan, p. 166.
51. For use of this tactic in organizations, see Wilson, pp. 211-213.
52. Horace Kallen, "Innovation," in Seligman and Johnson, Encyclopedia of the Social Sciences, reprinted in Etzioni, p. 429.
53. See Milton Rokeach, The Open and Closed Mind (New York, 1960) esp. pp. 55 and 286-288. Kenneth Jowitt makes imaginative and productive use of the "open-closed mind" distinction in his Revolutionary Breakthroughs and National Development: The Case of Romania (Berkeley: University of California Press, 1971, Forthcoming).
54. Vernon V. Aspaturian, The Soviet Union in the World Communist System (Stanford, 1966) p. 84.
55. "The degeneration of steering performance and learning capacity may be the direct consequence of survival and success themselves." Deutsch, Nerves of Government, p. 228.
56. See Zygmunt Bauman, "The Influence of East European Social Science on Soviet Social Science," Paper for the Conference on The Influences of Eastern Europe and Western Areas of the USSR on Soviet Society, Center for Russian and East European Studies, University of Michigan, Ann Arbor, May, 1970.
57. James N. Rosenau, "Toward the Study of National-International Linkages," in Rosenau, Linkage Politics, p. 45.
58. Shepard, p. 520.
59. "Deviation Control," pp. 16-18.
60. See "Letter from the Central Committee of the CPSU to Comrade Tito and other Members of the Central Committee of the Communist Party of Yugoslavia," etc. (27 March 1948), in Royal Institute of International Affairs, The Soviet-Yugoslav Dispute (London and New York 1948), pp. 15-63.
61. Adam Ulam, Titoism and the Cominform (Cambridge, Mass. 1952), pp. 108.
62. Joint Soviet-Yugoslav Declaration, Belgrade, June 2, 1955, in Robert Bass and Elizabeth Marbury, eds., The Soviet-Yugoslav Controversy 1948-1958 (New York, 1959), p. 57.
63. Pravda, November 23, 1956, quoted in ibid., pp. 78-81.
64. Kommunist, No. 6 (April), 1958, quoted in ibid., pp. 143-163. See also Khrushchev's speech at the Fifth Congress of the Socialist Unity party of Germany, in Pravda, July 12, 1958, pp. 2-3.
65. Vaclav Benes, Robert F. Byrnes, Nicholas Spulber, The Second Soviet-Yugoslav Dispute (Bloomington, Ind., 1959), p. xxx.

66. See Yuri Zhilin and Vadim Zagladin, "Yugoslavia Today," New Times, No. 34 (August 28), 1963, pp. 6-8, and the continuation of this article in No. 35 (September 4), pp. 3-5. See also V. Zagladin, A. Mitil, et al., "Yugoslavia Today—Journalists' Notes," World Marxist Review, Vol. VII, No. 3 (March, 1964).

67. See Natalia Sergeyeva and Irina Trofimova, "Yugoslavia: Economic Reform in Action," New Times No. 49 (December 6), 1967, and L. Tyagunenko, "Yugoslavia's Economic Reform," New Times No. 17 (April 28), 1968.

68. Yu Georgiyev, "Yugoslavia: 'New Variant of Socialism?' " Kommunist No. 15 (October), 1968, translation in Current Digest of the Soviet Press, Vol. XX, No. 52, January 1969, p. 3.

69. The literature on developments in Czechoslovakia is extensive and probably familiar to most readers. Perhaps the best single English language source is Robin Alison Remington, Winter in Prague (Cambridge, Mass., 1969). Another useful publication is Studies in Comparative Communism, Vol. I, Nos. 1 and 2 (July, October, 1968). For the most important official documents, see Rok Šedesaty Osmy (Praha, 1969). A good secondary analysis may be found in Philip Windsor and Adam Roberts, Czechoslovakia 1968 (New York, 1969).

70. Studies in Comparative Communism, pp. 294-295.

71. Based on F. Konstantinov, "Marxism-Leninism: A Unitary International Theory," Pravda, June 14, 1968; "The Train Jan Prochazka Missed," Literaturnaia Gazeta, May 8; I. Alexandrov, "Attack on the Socialist Foundations of Czechoslovakia," Pravda, July 11; The "Five-Party Letter" from the Bulgarian, East German, Hungarian, Polish, and Soviet Communist Parties to the Czechoslovak Party, July 14-15; "Defense of Socialism as the Highest International Duty," Pravda, August 22. For very similar statements by the East Germans, see On the Situation in the Czechoslovak Socialist Republic, np., n.d.

72. See, for example, Richard Lowenthal, "The Sparrow in the Cage," Problems of Communism, Vol. XVII, No. 6 (November-December, 1968), p. 10, and, especially, Radio Free Europe, "Pro-Czechoslovakian Mood in the Ukrainian SSR," July 16, 1968.

73. Peter J. Potishnyj and Grey Hodnett, The Ukraine and the Czechoslovak Crisis, Australian National University, Department of Political Science and Research School of Social Sciences, Occasional Paper No. 6, (1970).

74. Ibid., pp. 102, 115-116, 117.

75. Ibid., p. 75.

76. "While the Soviet leadership stood to lose from the exposure of the Ukrainian intelligentsia, youth, and public at large to greater knowledge of the Czechoslovak reforms, it probably stood to gain from the exposure of many Ukrainian officials to these threatening phenomena. Thus, 'public opinion' losses were to some extent offset by a gain in support for repressive action from within the Soviet political machine." Ibid., p. 120.

77. Ibid., pp. 122-123.

78. See A. Hegedus, "On the Alternatives of Social Development," Kortars, June, 1968, Radio Free Europe, Hungarian Press Survey, No. 1947 (September 13, 1968).

79. Harry G. Shaffer, "Progress in Hungary," Problems of Communism, Vol. XIX, No. 1 (January-February, 1970), pp. 49-50.

80. Matyas Toth, "Hungary: Strengthening the Leading Role of the Party," World Marxist Review, Vol. 12, No. 8 (August, 1969), p. 24.

81. L. Mosin, "Slavnyi Put' Borb'y za Mir i Sotsializm," Kommunist, No. 17, November, 1968, pp. 86-87.

82. Imre Pardi, "The Experience of the New System of Managing the National Economy in the Hungarian People's Republic," Ekonomicheskaia Gazeta, No. 46, November, 1968, translated in William F. Robinson, "Hungary's NEM: A Documentary of Soviet Views and Magyar Hopes," Radio Free Europe, May 30, 1969.

83. Shaffer, p. 51.

84. Michael Gamarnikow, Economic Reforms in Eastern Europe (Detroit, 1968), pp. 56-57. This book is useful for comparing the Czechoslovak and Hungarian economic reforms, as well as for comparison of East European economic reforms in general.

85. Népszabadsag, September 1, 1968, quoted in RFE "Hungarian Reform after the Invasion," October 14, 1968, p. 24.

86. Jozsef Bognar, "Economic Reform and International Economic Policy," The New Hungarian Quarterly, Vol. IX, No. 32 (Winter, 1968), p. 84.

87. Radio Free Europe, "Party Daily Urges Bolder Use of the Veto," June 6, 1969. On the new status of the unions, see Sandor Gaspar, "Increased Role and Tasks of the Trade Unions in Our Country," RFE Hungarian Press Survey No. 1745 (September 1966); RFE, "Characteristic Features of the New Hungarian Labor Code," January 15, 1967; RFE, "Hungarian Trade Union Activities: Where Politics and Economics Meet," January 31, 1969; and Gamarnikow, pp. 147-154.

88. Fred E. Dohrs, "Incentives in Communist Agriculture: The Hungarian Models," Slavic Review, Vol. XXVII, No. 1 (March, 1968), p. 25.

89. Ibid., p. 27.

90. RFE, " 'Socialist' Democracy on the Move: Kallai Outlines Parliament's Action Program for 1968," March 8, 1968, p. 2.

91. "Otvetsvennost' Pered Vremenem," Izvestiia, July 25, 1968, p. 5. See also Rezso Nyers' call for a further increase in the role of parliament in RFE, "L'Unita Interviews Rezso Nyers," July 22, 1969.

92. "Otvestvennost,' " p. 5.

93. Clyde Farnsworth, "Hungarians Vote for Parliament," New York Times, April 26, 1971.

94. Andrew Gyorgy, Nationalism in Eastern Europe, Report, Research Analysis Corporation, RAC-R-89, January, 1970, p. 46.

95. RFE, "The Literary Scene in Hungary," June 13, 1969.

96. Gamarnikow, p. 120.

97. Imre Vertes, "The Workers' Party, The Workers' Policy," Népszabadsag, May 1, 1969, quoted in RFE, "What is a Proletarian?" June 25, 1969.

98. Janos Gosztonyi, "The Party, Workers' Power, and Socialist Democracy," Pravda, June 1, 1969, quoted in RFE, ibid.

99. L. Rozsa, "The Road of our Democracy," Népszabadsag, September 1, 1968, RFE Hungarian Press Survey No. 1950, September 18, 1968. Barnabas Racz takes a dim view of Hungarian political change in "Political Changes in Hungary After the Soviet Invasion of Czechoslovakia," Slavic Review, Vol. XXIX, No. 4 (December, 1970).

100. Jeno Fock in Magyar Nemzet, September 26, 1968 and Népszabadsag, September 22, 1968, quoted in RFE, "Hungarian Reform after the Invasion," October 14, 1968.

101. Mosin, p. 87. See also Pravda, July 4, 1968 and the article by V. Gerasimov in Pravda, July 8, 1969. V. Gutsev and M. Petunin in Mezhdunarodnaia Zhizn' (No. 6, June 1968), emphasize Hungary's fidelity to the Warsaw Pact and Comecon, as did Brezhnev at the Tenth Party Congress in November, 1970.
102. Pravda, November 25, 1970.
103. Pardi in Robinson, p. 17.
104. Tibor Pethe, "Rovnyi Pul's Budapesht," Izvestiia, January 18, 1969.
105. B. Rodionov, "Verdni Sochinitelei iz 'Veltvokhi,' " Izvestiia, November 27, 1969.
106. See, for example, J. Horvath in Népszabadsag, December 7, 1970 (RFE Hungarian Press Survey, No. 2109, January 21, 1971) and Z. Lokosz in Magyar Hirlap, December 6, 1960 (RFE Hungarian Press Survey, no. 2108, December 30, 1970).
107. "Hungarian Reform after the Invasion," p. 9. See also the very important editorial, "Where are we Going?" Népszabadsag, September 22, 1968, RFE Hungarian Press Survey No. 1953, October 9, 1968.
108. RFE, "Hungarian and Soviet Reforms Compared: Nyers' Statement," September 16, 1969.
109. RFE, "Izvestiia Approves Hungarian Farm Council," March 25, 1968.
110. Dohrs, p. 36.
111. RFE, "Pravda Personalizes Nadudvar," August 19, 1968.
112. Handelsblatt, December 2, 1968, quoted in Robinson, p. 33.
113. Shaffer, p. 59.
114. Bernard Gwertzman, "Soviet Lag Stirs 'Self-Criticism,' " New York Times, March 26, 1970.
115. V. Savin, "Problemy i Perspektivy Sotsialiticheskoi Demokratii," Novyi Mir No. 5, May, 1969, p. 268.
116. Aryeh L. Unger, "Politinformator or Agitator: A Decision Blocked," Problems of Communism, Vol. XIX, No. 5 (September-October, 1970), p. 34.

ST. MARY'S COLLEGE OF MARYLAND
ST. MARY'S CITY, MARYLAND

058063

ZVI Y. GITEL[MAN] ...where he also taught, a[nd] ... [t]he University of Mic[higan] ...[Russi]an and East Europea[n] ... [a]rticles to Survey, Pr[oblems of Communism] ... "Power and Authori[ty in] ... [Com]munist Systems (19[ ]) ... [Pri]nceton University P[ress] ... [politi]cs.

JN 96 .A3 1972

058063

Gitelman, Zvi
The diffusion of political innovation
78

DATE DUE